study
eres

An econometric analysis of the household demand for electricity in Great Britain

An econometric analysis of the household demand for electricity in Great Britain

R. J. Ruffell

University of Stirling

1977

Scottish Academic Press

Edinburgh and London

Published by
Scottish Academic Press Ltd.
33 Montgomery Street, Edinburgh EH7 5JX

Distributed by
Chatto and Windus Ltd.
40 William IV Street
London WC2N 4DF

ISBN 0 7073 0140 8

Printed in Great Britain by
R. & R. Clark Ltd., Edinburgh

TO HELEN

PREFACE

It is my pleasure to acknowledge the generous help that I have received from many individuals and organizations at all stages of my work.

My thanks go especially to Professor J. A. C. Brown for originally suggesting the subject and subsequently providing support, encouragement, and valuable comments on the manuscript; and to:

Members of the staffs of the Electricity Council and South Western Electricity Board for freely giving of their time to answer my queries and allowing me thus to draw on their invaluable practical knowledge of the industry, and, in particular, to Mr T. A. Boley of the Electricity Council and Mr B. J. Rogers of SWEB.

The Social Science Research Council for financing me for three years.

Audits of Great Britain Limited for permission to use information from their sample surveys of appliance ownership.

The Electricity Council for permission to use their data on electricity consumption, tariffs, and appliance ownership.

The Gas Council for permission to extract details of gas tariffs from their files.

The Department of Employment for making available unpublished regional figures from the Family Expenditure Surveys.

The University of Stirling for providing the financial assistance which has made publication possible.

I am alone responsible for the way in which the data have been used and interpreted.

The computer programs were run on the University of Bristol's Elliott 503 and ICL 4-75 machines, the University of Stirling's Elliott 4130 and the Science Research Council's Atlas at Chilton, Berkshire. They were written in Elliott ALGOL, and the main programs also in ALGOL W for use on the ICL machine.

The first version of the manuscript was completed in December 1972 and was submitted to the University of Bristol as a doctoral dissertation. I have since taken the opportunity to undertake some revision, though I have not attempted to take account of improvements in the data available or developments in the literature. In doing so, I have benefited from the comments of Professors A. D. Bain, C. V. Brown, and G. R. Fisher. The responsibility for remaining deficiencies is entirely mine.

The main change from the dissertation is the deletion of much of the detail of the data processing. I should be happy to provide further detail to anyone who is interested.

The detail which remains is, I believe, essential to the argument and a proper understanding of the results. I also believe that no brief verbal summary of the algebraic exposition could be made without undue sacrifice of rigour. I have therefore not attempted to present the general reader with a separate summary but instead have indicated by asterisks in the List of Contents those sections which provide a guide to the aim of the study, its methodology, development, and results and the conclusions drawn.

Department of Economics,
University of Stirling
June 1974

CONTENTS

Preface   vii

1 INTRODUCTION
   *1.1 Introduction   1
   *1.2 Data   1
   *1.3 Aim   3
   *1.4 Use of quarterly time-series   4
   *1.5 Pooling of Area and time-series data   5
   *1.6 Classical theory   6
   *1.7 Modifications   6
   *1.8 Limitations   9

2 REVIEW OF THE LITERATURE
   *2.1 Introduction   11
    2.2 Stone (1954)   11
    2.3 Stone & Rowe (1958)   13
    2.4 Houthakker & Taylor (1966)   15
    2.5 Cramer (1959)   15
    2.6 Houthakker (1951)   17
    2.7 Doumenis (1965)   18
    2.8 Balestra (1967)   19
    2.9 The total demand for fuel   21
   2.10 Wigley (1968)   23
   2.11 Fisher & Kaysen (1962)   24
   *2.12 Conclusions   26

3 THE MODEL
   *3.1 Introduction   27
    3.2 The basic identity   27
    3.3 The demand equation for an individual consumer   28
    3.4 Aggregation over consumers   29
    3.5 Variables in the utilization functions   31
    3.6 The validity of the assumptions   33
    3.7 Modification to allow for the Z-U effects   36
    3.8 Modification to allow for variations in the availability of a gas supply   37
    3.9 The final form   39

3.10 The error term   40
3.11 Identification   40

4 CALCULATION OF THE VARIABLES

*4.1 Introduction   43
4.2 The definition of the dependent variable   43
4.3 The difference between units billed and units consumed   44
4.4 Aggregation over time   45
4.5 The error variance-covariance matrix   46
4.6 Calculation of the monthly weights, λ_q^m 46
4.7 Calculation of the dependent variable, Y^u 48
4.8 Ownership of space-heating and water-heating appliances—$Z^1, Z^2, Z^{o1} Z^{o2}$ 48
4.9 Ownership of cooking appliances—Z^3 49
4.10 Ownership of lighting and sundry appliances—Z^4, Z^{41}, Z^{42} 50
4.11 The reliability of the estimates of appliance ownership   51
4.12 The availability of a gas supply—G 52
4.13 Evidence on the relationship between ownership of electrical appliances   52
     and availability of a gas supply
4.14 The price of electricity—P^e 53
4.15 The price of gas—P^g 55
4.16 The price of solid fuel—P^c 57
4.17 Total consumption expenditure per household—E 57
4.18 The index of all other prices—π 60
4.19 Lagged variables   60
4.20 The natural variables—L, W 63

5 ESTIMATION AND TESTING: METHODS

*5.1 Introduction   67
5.2 Notation   67
5.3 The method of estimation   68
5.4 The method of computation   69
5.5 Test statistics, measures of reliability and goodness of fit   71
5.6 Analysis of the pattern of residuals   73
5.7 Testing for serial correlation   74
5.8 Testing for heteroscedasticity   76
5.9 Prediction   77
5.10 Invalidity of the assumptions   78

6 ESTIMATION AND TESTING: RESULTS

*6.1 Introduction   80
6.2 Parameters constrained to be zero   81
6.3 Estimation of the temperature threshold parameter, τ^u 84

6.4 Selection of the cooker ownership variable, Z^3, and lag distribution 85

6.5 Constraints on elasticities 89

6.6 The constrained demand equation 91

6.7 Goodness of fit 93

6.8 The pattern of the residuals 94

6.9 Estimated elasticities of unrestricted demand 95

6.10 The interpretation of the coefficients 96

6.11 Estimates of utilization 99

6.12 Prediction 101

6.13 Comparison with the results of previous studies 103

7 AN ANALYSIS OF DEMAND ON OFF-PEAK TARIFFS

*7.1 Introduction 106

7.2 The demand equation for an individual consumer 107

7.3 Aggregation over consumers 108

7.4 Variables in the utilization functions 108

7.5 Modification to allow for the Z-U effects 110

7.6 Modification to allow for variations in the availability of a gas supply 110

7.7 The final form 111

7.8 Aggregation over time 111

7.9 Calculation of the variables 112

7.10 Ownership of off-peak space-heaters and water-heaters—Z^{o1}, Z^{o2} 112

7.11 The quantity variable—Q^o 113

7.12 The price of off-peak electricity—P^{eo} 114

7.13 The temperature variable—W^o 114

7.14 Estimation and testing: methods 115

*7.15 The demand equation to be estimated 115

7.16 Estimation of the temperature threshold parameter, τ^o 116

7.17 Selection of the lag distribution 117

7.18 Constraints on elasticities 118

7.19 The constrained demand equation 118

7.20 Goodness of fit 119

7.21 The pattern of the residuals 122

7.22 The interpretation of the coefficients 122

7.23 Estimates of utilization 123

7.24 Prediction 125

8 A VARIANT MODEL WITH A TARIFF FUNCTION

*8.1 Introduction 127

8.2 The demand equation 128

8.3 The electricity price variable, P^{eu} 128

8.4 The aggregate tariff function: data 130

8.5 The aggregate tariff function: form 132
8.6 The reduced form 133
8.7 Estimation and testing: methods 134
8.8 Results 134
8.9 Omission of the price function 138
8.10 Conclusions 139

9 CONCLUSION
*9.1 Characteristics of the analysis 141
*9.2 The results achieved 142
*9.3 Future research 144

Appendix: List of notation 145
List of works cited 151
Statistical sources 152
Index 155

TABLES

1.3.1   Domestic electricity consumption by Area   3

1.3.2   The seasonal pattern of domestic electricity consumption in the South West   4

2.2.1   Stone (1954) results   12

2.3.1   Stone & Rowe (1958) results for fuel and light   14

2.6.1   Houthakker (1951) results   17

2.7.1   Doumenis (1965) results   19

2.10.1 Wigley (1968) results   24

4.6.1   The monthly weights, λ_q^m 47

6.3.1   Unrestricted demand: results with different estimates of the temperature   84
        threshold

6.4.1   Unrestricted demand: results with different cooker ownership variables   86
        and lag distributions

6.4.2   Zero-order correlation coefficients: ownership variables   87

6.4.3   Unrestricted demand: results with 26 variables, proportion owning as the   88
        cooker ownership variable, and lag weights w_1

6.6.1   Unrestricted demand: results with the constrained equation (6.6.1)   92

6.7.1   Consumption on unrestricted tariffs by Area   93

6.7.2   The seasonal pattern of unrestricted consumption in the South West   93

6.9.1   Unrestricted demand: β-elasticities estimated from equation (6.6.1)   96

6.12.1 Predictions of unrestricted demand in the South of Scotland from   102
        equation (6.6.1)

6.12.2 Annual consumption on unrestricted tariffs in the South of Scotland   102

6.13.1 Comparative analysis: results with equations (6.13.1) and (6.13.2)   104

7.10.1 The observations from which the off-peak equation was estimated   113

7.16.1 Off-peak demand: results with different estimates of the temperature   116
        threshold

7.17.1 Off-peak demand: results with 17 variables and lag weights w_1 118

7.19.1 Off-peak demand: results with the constrained equation (7.19.1)   120

7.19.2 Off-peak demand: β-elasticities estimated from equation (7.19.1)   120

7.20.1 Consumption on off-peak tariffs by Area   121

7.24.1 Predictions of off-peak demand in the South of Scotland from equation   125
        (7.19.1)

7.24.2 Annual consumption on off-peak tariffs in the South of Scotland   125

8.8.1   Tariff function model: results with the reduced form equation (8.6.1)   136

8.8.2   Results when the price function is omitted: comparison with the other   136
        results

8.10.1 Unrestricted demand: β-elasticities estimated from equation (8.6.1)   140

FIGURES

1.3.1. Domestic electricity consumption in England and Wales   2

6.8.1. The residuals, e, from estimating equation (6.6.1.): England and Wales   94
average

6.11.1. The estimates of the identifiable parts of the unrestricted utilization   100
functions for England and Wales from the results with equation (6.6.1)

7.23.1. The estimates of the identifiable parts of the off-peak utilization functions   124
for England and Wales from the results with equation (7.19.1)

8.8.1. The relative price of electricity in England and Wales   135

CHAPTER 1

INTRODUCTION

"[In] the analysis of market demand . . . we have essentially two questions: first, Is the ordinary demand theory, or some suitable modification of it, capable of explaining observed variations in the demand for different commodities?; second, if so, What are the most acceptable estimates of the demand parameters, particularly the income and price elasticities, and what is the reliabliity of these estimates?"

R. STONE (1951): *The Role of Measurement in Economics*

1.1 Introduction

This study is an attempt to answer these questions for electricity. In this chapter we introduce the theoretical and empirical propositions which form the basis of the model subsequently developed, in Chapter 3. In the interests of clarity and conciseness, the main description of the processing of the data to meet the requirements of the model is postponed until Chapter 4. Chronologically, the development of the model necessarily progressed side by side with the collection and processing of the data, since the exact form of the model could only be determined when the limitations imposed by data availability were known; and yet the collection and processing of the data could not be completed until the requirements of the model were worked out. In order, therefore, to indicate the influence on the model-building of this simultaneous work, we begin with a brief description of the data.

1.2 Data

Data were available for each of the twelve areas of England and Wales covered by the regional Electricity Boards on total domestic electricity consumption in kilowatt-hours, the number of domestic electricity consumers—and the proportion having a gas supply—, consumption on off-peak tariffs, the electricity tariffs, and the ownership of electrical appliances by type of appliance. Data on prices of solid fuels and temperature were available for individual places; on hours of daylight for different latitudes; on gas tariffs for each of the eleven areas covered by the regional Gas Boards; and on total consumption expenditure per household for the Standard Regions used in

government statistical sources. Both the Gas Board areas and Standard
Regions differed from the Electricity Board Areas. It was, however, possible
to derive figures referring to Electricity Board Areas for all the variables
mentioned.

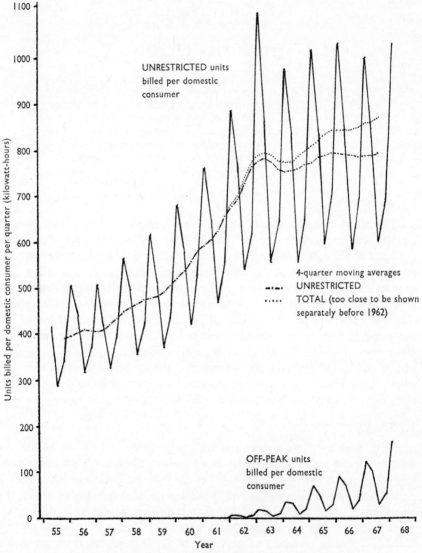

FIG. 1.3.1. Domestic electricity consumption in England and Wales.

Note: Units billed differ from units consumed in a quarter – see 4.3 below.

A quarterly time-series was available for domestic electricity consumption back to the second quarter of 1955. The latest figure available for this study referred to the first quarter of 1968. The dates of introduction of new electricity and gas tariffs were available for this period. Quarterly series for temperature and hours of daylight could be derived from data for shorter periods. The series for other variables were irregular, but the data were sufficiently spread out over the whole period of thirteen years to permit interpolation of the missing values. Thus series extending over 52 quarters for each of twelve areas were available in a form suitable for regression analysis.

1.3 Aim

The specific question which prompted us to embark on this study was, why has the domestic consumption of electricity been rising faster in the South West than in the country as a whole in recent years? As Table 1.3.1 shows, total consumption per consumer was growing at a rate of 4·8% per annum over the three years 1964/65 to 1967/68 in the South West, compared with a national average of 3·4%, and this rate was higher than in any other Area except South Wales.

TABLE 1.3.1. *Domestic electricity consumption by Area*

Electricity Board	Units billed per domestic consumer			
	1955/56	Growth rates		1967/68
		'55/56 to '61/62	'64/65 to '67/68	
	kWh	% p.a.	% p.a.	kWh
London	1617	8·1	1·9	3272
South Eastern	1901	7·2	3·1	3830
Southern	1711	9·9	3·4	4056
South Western	1586	8·4	4·8	3810
Eastern	1818	8·0	3·3	3960
East Midlands	1458	7·4	4·0	3160
Midlands	1646	8·7	3·1	3845
South Wales	1209	8·8	4·8	2898
Merseyside and North Wales	1483	8·9	3·5	3500
Yorkshire	1297	11·1	2·7	3326
North Eastern	1104	9·8	4·7	2731
North Western	1445	8·9	3·6	3389
England and Wales	1550	8·7	3·4	3528

Notes: 1 unit = 1 kilowatt-hour (kWh).
The years quoted are years beginning 1 April.

In attempting to answer this question, one is naturally led to the more general question, what determines variations between areas and over time? Consequently our aim may be stated thus: the formulation, estimation, and testing of a model to explain the variations in domestic electricity consumption over time and between areas. The sorts of variations to be explained are illustrated in Figure 1.3.1 and Tables 1.3.1 and 1.3.2.

TABLE 1.3.2. *The seasonal pattern of domestic electricity consumption in the South West*

Quarter	Total units billed per domestic consumer: ratio of South West figure to England and Wales mean: geometric means for the period 1955II-1968I
I	0·995
II	1·024
III	1·068
IV	1·048
Year	1·033

Figure 1.3.1 shows that there was a particularly strong upward trend in total consumption per consumer in the first seven years, 1955/56 to 1961/62. In the last four years, 1964/65 to 1967/68, the growth of total consumption was slower, and almost all the growth was in consumption on off-peak tariffs. (In the intervening two years, the annual figures were exceptionally high owing to the severity of the 1962/63 winter.) The marked seasonal pattern displayed irregularities from year to year.

Table 1.3.1 shows that between Areas there were large differences in consumption levels at all dates, and in growth rates; and the rankings of Areas in terms of both changed over the period. There were also variations in the seasonal pattern: Table 1.3.2 shows that, in the South West for example, consumption per consumer was on average 3·3% higher than in England and Wales as a whole, but in the third quarter 6·8% higher and in the first quarter 0·5% lower, that is, in the South West consumption varied less between winter and summer.

1.4 Use of quarterly time-series

In all the analyses, quarterly time-series have been used. The disadvantage of this is that, for the total consumption expenditure and appliance ownership variables, many interpolated values have to be used. Hence we may expect errors in these variables of a systematic kind, resulting in biased and inconsistent estimators, higher correlation between the explanatory variables, and therefore less reliable estimators. Two circumstances, however, tend to reduce the seriousness of these deficiencies: (1) total consumption expenditure seasonally adjusted, and appliance ownership would have changed fairly gradually from one quarter to the next, so that errors due to interpolation will be small; (2) the correlation between the errors and the true values of a variable will be low.

The advantages outweigh this disadvantage. Improved estimates of the coefficients of the price variables and the natural variables—temperature and hours of daylight—may be obtained by using quarterly instead of annual data: more accurate estimation of lagged price effects is possible, given that changes in prices occurred at different dates during the year; the shorter the time-period used in estimating non-linear relationships, the less unsatisfactory is the use of arithmetic means of daily or monthly figures.

We have not attempted to explain variations before 1955, for the following reasons: (1) it is unlikely that the same model structure would apply because of the effects of rationing and other post-war controls; (2) only annual data were available on electricity consumption in earlier years; and (3) it was possible to extend the series for explanatory variables back to before 1955 to allow the inclusion of lagged variables in the model without reducing the number of observations, but lagged variables could not have been included if the data on electricity consumption before 1955 had been used.

1.5 Pooling of Area and time-series data

In all the analyses, the Area and time-series data have been pooled. Few coefficients could be satisfactorily estimated from any one time-series of 52 quarterly values because of the high correlation over time between many of the explanatory variables. One set of twelve Area observations does not provide enough degrees of freedom, given the large number of variables which one would expect to make a significant contribution to the explanatory power of the model and on which data were available. Pooling allows the model to be estimated with relatively few restrictions on parameters, by providing a large number of observations and reducing the correlation between explanatory variables.[1]

Two assumptions are implicit in the use of pooled data:

Assumption 1.5.1

There are no variations in parameters between Areas, and no variables which cause systematic variations between Areas are omitted.

The validity of this assumption will be examined in Chapters 6 and 7.

(1) See Balestra (1967) pp. 69-70, 136 and Bain (1964) p. 51.

Assumption 1.5.2

The structural parameters are the same whether the variations in the variables are over time or between Areas. This implies correct specification of the model with respect to the variables included, in particular lagged variables.

The validity of this assumption will be examined in Chapter 4.

Some data were available for the South of Scotland Electricity Board for the latter half of the period. This has been used for a predictive test of the model.

1.6 Classical theory

Our starting-point in formulating the model was the classical theory of consumers' behaviour according to which a consumer maximizes his utility subject to a budget constraint. It is assumed that the number of commodities, say N, is fixed; that tastes remain constant; and that income and prices are given independently of the quantities purchased. On the usual assumptions about the consumer's preferences,[2] it can be shown that the quantity demanded of each good by the consumer may be expressed as a function of the N prices and income, and that a number of restrictions on the coefficients of these demand functions are implied:

   (i) the demand functions are single-valued;
   (ii) they are homogeneous of degree zero in prices and income;
   (iii) the aggregate effect of reallocations of the budget due to income and price changes is that income continues to be exhausted;
   (iv) the substitution matrix, the matrix of price derivatives when income is set so that utility is left unaltered after a price change, is symmetric and negative semi-definite.

Given no change in the distribution of income and certain weak conditions on the individual Engel curves, these results carry over to market demand.[3]

1.7 Modifications

The simplest model would comprise a single equation expressing the quantity demanded of electricity as a function of its current relative price and real income.[4] Restriction (i) could be applied by choice of a suitable functional form, e.g. the linear, semi-log, or log-linear, and restriction (ii) by deflating the current price of electricity by an index of all other prices. The other theoretical restrictions and empirical evidence suggest a number of ways in which this model could be improved:

(2) See Wold (1953) ch. 4, Samuelson (1947) ch. 5.
(3) See Stone (1954) ch. 18, Brown & Deaton (1972) section II, Pearce (1964) ch. 3.
(4) See, e.g., Stone & Croft-Murray (1959) p. 55 et seq.

(1) The theory leads to a specification which includes a set of N simultaneous equations, one for each commodity. The quantity demanded of the ith commodity is expressed as a function of all N prices and income; restrictions (i) and (ii) apply to each function and restrictions (iii) and (iv) to the matrix of coefficients. The enormity of the task precludes estimation of N simultaneous equations in N prices and income. But a fairly well defined division of the N commodities suggests itself: there is a fairly distinct set of wants satisfied by fuels and fuel-using appliances; many of the wants satisfied by electricity may be satisfied by other fuels, but commodities other than fuels are relatively poor substitutes. Hence commodities may be divided into two groups: electricity, other fuels and fuel-using appliances; and all other goods. Estimation of the demand equation for electricity is likely to be improved by simultaneous estimation of equations for as many commodities as possible from the first group, but relatively little affected by the exclusion of equations for commodities in the latter group. Similarly the demand equations for the fuels should include as many fuel and appliance prices as possible but all other prices may be adequately represented by a single index number.

(2) The theory assumes that the tastes of "a consumer" remain constant. But the single consumer of theory is an economic decision-making unit, a household whose size and composition, and therefore wants, change. There may also be variations in the wants of a household apart from those brought about by changes in its size or composition, for example in the desired room temperature. Therefore variables representing household size and composition and changing wants should be included in the demand equation.

(3) Even if wants in the ordinary sense remain constant, tastes in the sense required by theory will not.[5] The theory assumes that tastes remain constant in the sense required for the result that, if all prices and income remain constant, then so will the quantities demanded of each commodity. Tastes for electricity will not remain constant in this sense for two reasons:

(i) Some of the wants satisfied by electricity are also satisfied naturally to an extent that varies over time and between areas. Hence even if wants, e.g. for space-heating, remain constant, tastes in the theoretical sense will change, and demand shift, in response to changes in natural conditions. This should be allowed for by including natural variables in the demand equation.

(ii) The efficiency of electrical appliances may improve so that the same wants are satisfied with a lower consumption of electricity. For example, an improvement in the insulation of ovens would mean that a want for 400 °F. for one hour would be satisfied by using less electricity. This implies that some measure of the technical efficiency of each type of appliance should be included in the demand equation.

(5) On wants and commodities in utility theory, see Ironmonger (1972).

(4) The theory assumes supply to be perfectly price-elastic. Since, in the short run, the electricity tariff structure gave a step supply function for the individual consumer, and in the long run there were changes in the tariff which may have been related to shifts in demand, the simultaneous system of equations should be extended to include long- and short-run supply functions.

(5) Electricity satisfies many wants. There is no reason for supposing that the sensitivity to changes in prices, income, and tastes is the same for each use. Indeed the differences in the availability of substitute other-fuel appliances between uses suggest that the sensitivities are quite different. Since the composition of the total appliance stock will change over time, the demand for electricity should be expressed as a sum of demand functions for each use.

(6) The theoretical system of equations does not use the information that a necessary condition for electricity demand to be positive is possession of an electrical appliance. To allow for this particular form of complementarity, the demand function for each appliance type, or use, should include the appliance ownership level, and the form of the function be such that, if appliance ownership is zero, then demand is zero.

(7) Changes in the demand for electricity in response to changes in prices or income will depend on adjustments to the stock of appliances, and in utilization habits. As neither will be instantaneous, adjustment to a new equilibrium will take time. Therefore lagged values should be included in the demand equation, especially as quarterly and not annual data are used.[6]

(8) The demand for electricity during a quarter is a sum of the demands[7] at different hours, each of which will in general be a different function of prices, income and other variables, depending on the time of day, day of week, and whether the day is a holiday or working day. Since the proportion of hours with any given demand function does not in general remain constant from quarter to quarter, it would lead to an improvement in estimates of parameters to estimate as many separate demand functions as possible. In particular the model would be improved by separate estimation of the coefficients of natural variables in different hourly demand functions because maximum hourly demand is an important determinant of the supply price; and of the coefficients of the price of electricity because Electricity Boards introduced tariffs on which this price varied with the time of day.

(6) See Malinvaud (1966) p. 473.
(7) Throughout this study we use "demand" in the general sense of economics to mean "demand per unit time" measured in kilowatt-hours, the unit of time depending on the context. This meaning includes as a special case the one used by the electricity industry, viz. demand at an instant in time, given in kilowatts, usually as an average over half an hour.

1.8 Limitations

The available data limit the scope for incorporating these eight modifications in the model. The effect of the omissions on the properties and interpretation of the estimates obtained will be examined in Chapters 3 and 4.

(1) Simultaneous equations. Even a model confined to demand equations for fuels and fuel-using appliances requires a large amount of data for its estimation. The data limit the choice of model to two sorts: either one which includes a very simple equation for electricity and similar equations for some other fuels and appliances; or one which comprises a single equation for electricity of a relatively elaborate form. A conclusion on which of these to use will be reached after reviewing the results of previous studies in Chapter 2.

(2) Changes in household wants. Because of lack of sufficiently detailed data, no variables representing household size and composition or changing wants will be included in the model.

(3) Natural variables and technical coefficients. Data on the natural variables are relatively abundant; variables representing temperature and hours of daylight can be included. In the absence of any information, technical efficiency variables will have to be omitted.

(4) Supply. Although the current rates on the domestic tariff may be expected to have depended on the quantity currently supplied to domestic consumers, the relationship was probably a weak one, for reasons to be given in 3.11. The long-run supply function will therefore be omitted as acceptable estimates of the demand parameters may still be obtained and the need for an analysis of the other determinants of supply, such as the costs of generation, is removed.

In the main model developed in Chapter 3, every consumer in any given Area will be assumed to have the same marginal price; data on the structure of tariffs enables us to develop a variant model in Chapter 8 in which this assumption is relaxed.

(5) Disaggregation by type of appliance. The available data on appliance ownership permit disaggregation into four classes, called in the sources "space-heating", "water-heating", "cooking", and "lighting and sundry", and separation out of appliances on off-peak tariffs, for which separate consumption data are also available.

(6) Complementarity. This restriction is easily applied by introducing the calculated appliance ownership variables multiplicatively as explanatory variables.

(7) Lags. The availability of time-series data readily permits the inclusion of lagged values of any of the variables.

(8) Disaggregation over time. The only disaggregation of quarterly consumption on which we have data is into consumption on off-peak tariffs and

consumption on other, that is "unrestricted", tariffs. Now during the period under study up to 1968ı, the off-peak rates[8] applied to consumption during the night, and usually part of the afternoon as well, by specified appliances only—almost all being space-heaters of the indirect-acting floor-warming, ducted air, and storage types and immersion and storage water-heaters. Hence consumption on off-peak tariffs did not equal total consumption during the off-peak hours, or the total consumption of off-peak consumers during these hours.

We may regard an appliance connected to an off-peak meter as being of an identifiably different type from a physically identical one connected to an unrestricted meter. Therefore the division of quarterly consumption into two parts represents a disaggregation by type of appliance and only for those types connected to off-peak meters a disaggregation over time.

The analysis of demand on off-peak tariffs will be set out separately in Chapter 7. In the chapters on unrestricted demand (3 to 6 and 8) terms such as demand and consumption refer to that part on unrestricted tariffs unless we explicitly state otherwise.

[8] Except the night rate on the Eastern Electricity Board's "night and day" tariff. We use the term "off-peak" to denote any rate available only during restricted hours—see 7.1.

CHAPTER 2

REVIEW OF THE LITERATURE

2.1 Introduction

The models used in previous studies of the household demand for electricity have differed in respect of the number of the theoretical restrictions and modifications embodied in them. A review of them can therefore throw some light on the consequences of omitting these restrictions and modifications, on the merits of different ways of incorporating them, and on the data required. In the following sections we first examine each model in turn, starting with the simpler ones, and finally draw conclusions on the form of model which should be employed in this study. The works reviewed include several which analyse the demand for fuels other than electricity. They also refer to different time-periods and areas—Great Britain, the United Kingdom, or the United States; in part, the associated differences in market structure account for the differences in models and results, but some general conclusions can be drawn.

To facilitate comparisons we have used a standardized notation throughout, conforming as much as possible with that used in the rest of this study. It consequently differs from the notation used in all the works cited. For terms such as "income" and "the price of electricity", we have used the same letters throughout, and the same coefficient subscripts, although the exact definitions vary between authors. Our nomenclature is set out in the Appendix.

2.2 Stone (1954)[1]

Stone performed separate analyses of cross-section and time-series data. In both, simple static models were used which only differed from the simple model described in 1.6 by the inclusion of natural and household wants variables. The demand equation estimated from the cross-section data, on twelve working-class groups in 1937-38 and four middle-class groups in 1938-39 in the United Kingdom, was

$$\log (V/H) = \beta_1' + \beta_7' \log (E/H) + \beta_{12}' H + \beta_{13}' d + \varepsilon \qquad (2.2.1)$$

(1) pp. 222-237, 399-404.

where V is average expenditure per household on electricity,

H the average number of equivalent adults per household,

E average expenditure per household on all commodities "excluding National Health payments, pensions, unemployment insurance, insurance premiums, and payments to pension funds etc." (p. 313),

d a dummy variable, $=0$ for the working-class groups,
$=1$ for the middle-class groups.

The equation estimated (after taking first differences) from the time-series data for 1920-38 in Great Britain was

$$\log(Q/C) = \beta_1 + \beta_3 \log W + \beta_7 \log I^e + \beta_8 \log(P^e/\pi) + \beta_{14} t + v \quad (2.2.2)$$

where Q is the total quantity of electricity consumed,

C the number of electricity consumers,

W average temperature over the six months October to March,

I^e the average real disposable income of electricity consumers,

P^e the average price of electricity,

π an all-commodities price index,

t a linear time trend.

The principal results are set out in Table 2.2.1.

TABLE 2.2.1. *Stone (1954) results*

Source: Stone (1954) p. 400

Cross-section

Parameter	β_7'	β_{12}'	β_{13}'	$1 - \beta_7' + \beta_{12}'$	
Estimate	0·99	(1)	0·19	0·01	$R^2 = 0.92$
Standard error	0·22		0·21	0·22	

Time-Series

Parameter	β_3	β_7	β_8	β_{14}	
Estimate	0·19	0·19	−0·58	0·001	$R^2 = 0.78$
Standard error	0·21	0·12	0·15	0·004	Durbin-Watson $d = 1.58$

(1) Variable omitted because of high correlation with other variables

The cross-section results show that total consumption expenditure was an important determinant of demand. Since current and lagged values would be highly correlated in cross-section we cannot conclude, however, that only current expenditure should be included.

The influence of household size and composition, indicated by $1 - b_7' + b_{12}'$, appears to be negligible but, as Stone's equivalent adult scale was based on food expenditure, little confidence can be placed in this result.

In the equation estimated from the time-series, the positive but insignificant temperature coefficient is probably explained by (1) the inclusion of temperatures for only the six winter months, and (2) the predominance in the appliance stock during this period of types which were substituted for other-fuel appliances during the summer. Different results could be expected with a different composition of the stock.

Using the Houthakker (1952) result, quoted by Stone (p. 312), that the elasticity of total expenditure with respect to income for households in the 1938-39 survey was 0·9, the estimate of the income elasticity from the cross-section analysis is 0·89, whereas the estimate from the time-series analysis is only 0·19. Two features of the model probably account for most of this discrepancy: (1) because only current income enters the time-series equation and first differences are taken, the equation gives a short-run elasticity whereas the cross-section estimate is a long-run one; (2) the use of the average price of electricity without extension of the model to include a simultaneous equation specifying the dependence of average price on the quantity supplied produces a downward bias in the income elasticity b_7—and in the estimate of the price elasticity, b_8. On the other hand the omission of the average price of electricity from the cross-section equation means that the expenditure elasticity b_7 is upward biased.

2.3 Stone & Rowe (1958)

Stone & Rowe extended this model to allow for dynamic adjustment to equilibria. They first developed a model applicable to the demand for durables, based on three postulates:

(1) Consumption, or depreciation, Γ, is a constant proportion of the available stock at any time:

$$\Gamma = \phi Z + \mu Q, \qquad 0 \leqslant \mu \leqslant \phi \leqslant 1 \tag{2.3.1}$$

where Z is the stock at the beginning of a period,

$\quad\quad Q$ the quantity purchased during the period (assumed to be spread evenly over the period).

(2) The average level of stocks in a period, $\bar{Z}$, adjusts to the desired level, Z^*, according to the relationship

$$\frac{\bar{Z}}{Z} = \left(\frac{Z^*}{Z}\right) \tag{2.3.2}$$

(3) The desired level is given by

$$\ln Z^* = \beta_1 + \beta_7 \ln (E/\pi) + \beta_8 \ln (p/\pi) + \beta_{14} t + v \tag{2.3.3}$$

where E is total consumption expenditure,

$\quad\quad \pi$ an all-commodities price index,

$\quad\quad p$ the price of the commodity,

$\quad\quad t$ a linear time trend.

Perishable goods are treated as a limiting case: in (2.3.1) $\phi = \mu = 1$; and in (2.3.2) and (2.3.3) the stock variables are replaced by corresponding consumption variables. To estimate the demand for fuel and light from quarterly data they assumed that actual consumption also depended on daily mean air temperature (W) and daily mean hours of sunshine (L), and was subject to a regular seasonal pattern. Combining the equations and substituting expenditure at constant prices, V, for quantity purchased yields

$$\ln V_t = \beta_1 \theta + \beta_3 W_t + \beta_4 L_t + \beta_7 \theta \ln (E/\pi)_t + \beta_8 \theta \ln (p/\pi)_t$$
$$+ \beta_{14} \theta t + \beta_{15} d_1 + \beta_{16} d_2 + \beta_{17} d_3 + (1 - \theta) \ln V_{t-1} + \varepsilon$$

$$(2.3.4)$$

where d_1, d_2, d_3 are dummy seasonal constants.

This was estimated, after removing seasonal means and taking first differences, from data for the United Kingdom 1950-56, giving the results in Table 2.3.1.

TABLE 2.3.1. *Stone & Rowe (1958) results for fuel and light*
Source: Stone & Rowe (1958) p. 263

Parameter	β_3	β_4	$\beta \theta$	$\beta_8 \theta$	$1 - \theta$	
Estimate	-0.76	[1]	-0.06	-0.66	-0.10	$R^2 = 0.60$
Standard error	0.33		0.44	0.33	0.18	Durbin-Watson
						$d = 2.06$

(1) The coefficient of daily mean hours of sunshine was found to be insignificant. The only results published were for a regression omitting it.

We conclude that hours of sunshine are not a good measure of illumination or temperature; and that variations in temperature are an important part of the explanation of irregular seasonal variations in fuel consumption.

That the estimated value of θ does not differ significantly from one indicates that adjustment to a new equilibrium takes place within the same quarter. This is difficult to believe since it depends on adjustment to the stock of durables, which is slow according to Stone & Rowe's other results. If, however, there was a fairly strong trend in expenditure at constant prices, the time trend, t, would tend to take up the explanation not only of movements in the equilibrium but also of movements towards it, especially as the irregular seasonal variations in V_{t-1} were not perfectly explained by the temperature variable.

The results do not compare favourably with those obtained for other goods, particularly durables, or with the Stone (1954) results for fuel and light obtained from a static model. This suggests that some other dynamic specification should be tried for fuels.

2.4 Houthakker & Taylor (1966)

The influence of habit is introduced in a different way in Houthakker & Taylor (1966).

If, instead of the discrete time-periods considered by Stone & Rowe, continuous time is considered, equation (2.3.1) becomes

$$\Gamma = \phi Z \qquad (2.4.1)$$

If the adjustment process is assumed to depend on the difference between, instead of the ratio of, desired and actual stocks, then (2.3.2) is replaced by

$$\frac{dZ}{dt} = \theta(Z^* - Z) \qquad (2.4.2)$$

If desired stocks are assumed to depend linearly on price and income, but not to be subject to a time trend, then (2.3.3) is replaced by

$$Z^* = \beta_1 + \beta_7(E/\pi) + \beta_8(p/\pi) + v \qquad (2.4.3)$$

Using a linear approximation to the path taken by actual stocks within years, the following expression for the quantity consumed in year t may be derived:

$$Q_t = \alpha_1 + \alpha_2 Q_{t-1} + \alpha_3 \Delta'(E/\pi) + \alpha_4(E/\pi)_{t-1}$$
$$+ \alpha_5 \Delta'(p/\pi) + \alpha_6(p/\pi)_{t-1} + \varepsilon \qquad (2.4.4)$$

This is the demand equation obtained—by a different route—by Houthakker & Taylor, who then applied it to the estimation of the household demand for electricity in the United States. To apply this stock adjustment model to a perishable good they interpreted Z as a "psychological stock" of habits (p. 9). ϕ and θ are then parameters reflecting the speed of adjustment to a new equilibrium rate of consumption.

When the equation has the general form given, Ordinary Least Squares estimators are not optimal as it includes the lagged endogenous variable, Q_{t-1}, and is overidentified: $\alpha_3\alpha_6 = \alpha_4\alpha_5$. Applying it to their electricity data, Houthakker & Taylor found that the estimate of α_2 was close to one so that they were able to simplify estimation by using $\Delta'Q$ as dependent variable. However $\alpha_2 = 1$ implies $\theta = 0$ so that interpretation based on the stock adjustment model breaks down.

2.5 Cramer (1959)

The model used by Cramer in his analysis of the demand for petrol and oil from private motorists was static, but recognized that "the structure of demand is determined by the fact that non-motorists cannot consume petrol, while motorists are to some extent compelled to do so" (p. 334). To apply it to the explanation of variations in fuel expenditure between income classes using only information on the proportion of households in each class having the use of a motor vehicle and their fuel expenditure, three assumptions were made.

(1) Each household has a tolerance income, E^*,[2] such that it will only become a motorist household if its actual current income, E, exceeds E^*. The distribution of tastes is such that the tolerance incomes are distributed lognormally with mean μ and variance ω^2. Hence the proportion of motorist households at a given income level, $A(E)$, is given by

$$A(E) = \Lambda(E^* \mid \mu, \omega^2) \tag{2.5.1}$$

(2) The value of fuel consumed at constant prices by the ith household, V_i, depends only on its current and tolerance incomes:

$$
\begin{aligned}
V_i &= 0, & E_i &< E_i^* \\
V_i &= \theta_i + \beta_7(E_i - E_i^*), & E_i &\geqslant E_i^*
\end{aligned}
\tag{2.5.2}
$$

(3) The minimum consumption level, θ_i, is inversely related to the tolerance income level:

$$\theta_i = \kappa E_i^{*-\alpha} \qquad \kappa, \; \alpha \geqslant 0 \tag{2.5.3}$$

Hence the mean fuel expenditure by all households with income E is

$$\overline{V}(E) = \int_0^\infty V(E, E^*) d\Lambda(E^* \mid \mu, \omega^2) \tag{2.5.4}$$

and by motorist households only

$$\overline{V}_A(E) = \overline{V}(E)/A(E) \tag{2.5.5}$$

Given data on $A(E)$, μ and ω may be estimated from (2.5.1), and these estimates then used in (2.5.5) to estimate κ, α, and β_7 by an iterative method.

Good fits were obtained using data from two budget surveys in the United Kingdom in 1953-54. The simpler model with $\alpha = 0$ was found to be appropriate.

The model embodies the hypotheses (i) that a given variation in the proportion of households using motor vehicles is associated with different changes in fuel demand, depending on whether it is due to changes in tastes, as represented by tolerance incomes, or in income; and (ii) that marginal motorist households have below average demand. Cramer found support for these hypotheses in the slow growth of mean consumption per motorist household over the period 1950-57.

(2) See also Aitchison & Brown (1957), section 12.4, "The derivation of an Engel curve for an indivisible good."
    In fact total expenditure was used by Cramer, but we follow him in stating the theory in terms of income.

The model seeks to explain variations in ownership of the appliance as well as in expenditure on the fuel, i.e. ownership is an endogenous variable. This means that, to apply it to aggregate time-series data for all income classes, Cramer had to introduce the additional assumption that actual incomes, E_i, were lognormally distributed.

2.6 Houthakker (1951)

In the Houthakker (1951) model, the appliance stock is an exogenous variable, and lagged prices are introduced:

$$\log (Q/C)=\beta_1+\beta_2 \log Z+\beta_7 \log I+\beta_8 \log P_{t-2}^{ef}+\beta_{10} \log P_{t-2}^{g}+\varepsilon \quad (2.6.1)$$

where Q is domestic electricity consumption, in kilowatt-hours,
      C the number of electricity consumers,
      Z the average installed load, in kilowatts, per consumer of cookers, water-heaters and wash-boilers bought on hire purchase,
      I household income after tax per consumer,
      P_{t-2}^{ef} the marginal price of electricity, lagged two years,
      P_{t-2}^{g} the marginal price of gas, lagged two years.

(All variables refer to domestic consumers on two-part tariffs only.)

As Houthakker was only considering two-part tariffs, on which the same marginal rate applied to all consumers in a given town irrespective of their consumption, it was appropriate to use this marginal rate as the electricity price variable.[3] By a similar argument the marginal rate was chosen for the gas price. This required more data than the usually published figures since these give average prices. As the marginal rate was in the short run independent of the quantity demanded there was no need to introduce a function defining the tariff to obtain consistent estimators. No long-run supply function was included as he argued (p. 361) that, in the cross-section, supply was independent of the variables included in the equation.

The results of estimating the equation from a cross-section of 42 towns in Great Britain in 1937-38 are shown in Table 2.6.1.

The interpretation of estimates from models of this form depends on the relationship between the appliance ownership variable, Z, and the other

TABLE 2.6.1. *Houthakker (1951) results*
Source: Houthakker (1951) p. 367

Parameter	β_2	β_7	β_8	β_{10}	
Estimate	0·1767	1·166	−0·8928	0·2107	$R_2 = 0·87$
Standard error	0·0328	0·088	0·1905	0·1165	

(3) See 4.14.

variables included. As Houthakker found that the correlations were very low in his sample, the price and income elasticities may be interpreted as short-run ones indicating the effect on the utilization of a constant appliance stock. The cross-elasticity with respect to the price of gas is plausibly low given that in the short run substitution is possible only for those consumers who have both gas and electrical appliances. As the appliance stock is not dis-aggregated, the elasticities are averages of those for each type of appliance and are therefore only useful for the explanation of particular cross-section differences or prediction if the composition of the appliance stock was invariant.[4]

The low estimate of β_2 could simply mean that there was a negative cor-relation between the two parts of total installed load, those appliances in-cluded in Z and the remainder. If the remainder was constant as assumed by Houthakker, then an estimate of β_2 less than one indicates that marginal utilization was less than the average for cross-section differences in the appliance stock.

P_{t-4}^{ef}, P_t^{ef}, and P_t^g were also considered but the two-year lags were found to give the best results. The definition of I was such that it represented the long-run level of income. These results support the hypothesis that, quite apart from the effect of stock adjustment on the time taken to reach a new equilibrium, there is a lagged effect due to utilization habits.

Houthakker went on to analyse the monthly series for total electricity generated 1927-44. Although this refers to consumption by all users, the two main findings should apply to domestic consumption alone: the monthly pattern was almost entirely explained by two variables—hours of daylight and temperature; the sensitivity to temperature changes differed from month to month indicating a non-linear relationship with high sensitivity at winter temperatures but negligible sensitivity at summer temperatures.

2.7 Doumenis (1965)

A model which includes appliance ownership in the same way is estimated from time-series data in Doumenis (1965). The demand equation for domestic electricity consumption is

$$\log (Q/C) = \beta_1 + \beta_2 \log Z + \beta_7 \log (I/\pi) + \beta_8 \log (P^e/\pi) + \varepsilon \quad (2.7.1)$$

where Q/C is consumption per consumer,
  I personal income per head,
  π an all-commodities price index,
  P^e the average price of electricity,
  Z the appliance stock per consumer, in kilowatts.
The results are given in Table 2.7.1.

  (4) See also 2.11 below.

TABLE 2.7.1. *Doumenis (1965) results*

Source: Doumenis (1965) p. 128

Parameter	β_2	β_7	β_8	
Estimate	0·4512	1·328	−0·3719	$R^2 = 0.990$
Standard error		not given		

The time-series data required for this model are readily available—Doumenis used annual data for 1947-61. But there are two major problems in estimating the model from such data: firstly, the use of an average price for electricity leads to bias, as in the Stone (1954) time-series model; secondly, whilst good fits can be obtained, the estimates are unreliable because of the high correlations between the variables included and between these and omitted taste, lagged price and lagged income variables.

Doumenis went on to estimate a set of demand equations for different types of appliance, but this was done independently and no restrictions on the electricity equation were introduced.

2.8 Balestra (1967)

Balestra started by postulating a static model in his analysis of the demand for natural gas in the United States:

$$\log (J^g/N) = \beta_1 + \beta_7 \log (I/\pi) + \beta_{10} \log (P^g/\pi) + \varepsilon \qquad (2.8.1)$$

where J^g is the total gas consumption, in British Thermal Units,

   N the population,

   I income per head,

   π an all-commodities price index,

   P^g the price of gas (whether average or marginal is not clear).

J^g is adjusted for changes in the weather over time.

Expressing fuel consumption and income per head instead of per consumer is equivalent to expressing both per consumer and including mean household size as a separate variable with its parameter constrained to be $1 - \beta_7$.

The equation was estimated separately for each of 36 states, from annual time-series. Whilst the results were good in terms of fit, the estimates of individual parameters were regarded as unsatisfactory, b_{10} being generally positive and b_7 very high, because of correlation with omitted variables.

Balestra next considered a model which incorporated the complementarity modification. This model was developed from the identity

$$J^g \equiv Z^g U^g \qquad (2.8.2)$$

where Z^g is the total stock of gas appliances per head,

   U^g their utilization in hours.

A model based on an identity of this form is also used in Fisher & Kaysen (1962); the general implications will be discussed in 2.11.

It was postulated that

$$U^g = \beta_1 + \beta_7 \log (I/\pi) + \beta_{10} \log (P^g/\pi) + v \qquad (2.8.3)$$

and
$$Z_t^g = (1+\kappa)Z_{t-1}^g \qquad (2.8.4)$$

and hence that

$$\Delta' \log J^g = \log (1+\kappa) + \beta_7 \Delta' \log (I/\pi) + \beta_{10} \Delta' \log (P^g/\pi) + \varepsilon \qquad (2.8.5)$$

This only differs from (2.8.1) in the allowance for a trend and is formally equivalent to the Stone (1954) time-series model (except that J^g is adjusted for temperature changes). However, unlike Stone's results, the results were very poor judged by the values of the multiple correlation coefficient and Durbin-Watson d. Balestra argued that this was due to the exclusion of dynamic effects. These could be allowed for in two ways: (1) by including lagged values in the expression for U^g; or (2) by dropping the assumption of a constant rate of growth of the stock of appliances, Z. This requires either that Z be included explicitly in the estimated equation (which Balestra could not do because he did not have data on it), or that κ be respecified as a function of other variables.

In developing his dynamic model, Balestra introduced the concept of the "total demand for fuel", the implications of which we shall consider in the next section. The model also introduces the complementarity modification without using any data on appliance ownership, but several simplifications are made:

(1) The stock of all fuel-using appliances, Z_t^J,[5] is assumed to depend only on current real income per head and population, N:

$$Z_t^J = \beta_1 + \beta_7(I/\pi)_t + \beta_{18}N_t \qquad (2.8.6)$$

(2) The utilization of this stock is assumed constant:

$$U_t^J = U^J \text{ all } t \qquad (2.8.7)$$

(3) In terms of the Stone & Rowe (1958) model, it is assumed that (i) all adjustment to a new equilibrium level of Z_t^J takes place within the same year, i.e. $\theta = 1$ in (2.3.2); (ii) the depreciation rate is equal to the scrapping rate, i.e. scrapping is the only way in which appliances are consumed, and appliances are not scrapped in the period in which they are acquired: in (2.3.1), $\mu = 0$ and ϕ is the constant scrapping rate.

(4) The above three conditions are assumed to apply to the stock of gas appliances as well. They are scrapped at a rate ϕ^g.

The new demand for gas in year t, that is, the demand arising from the use of appliances newly acquired in year t, is then

(5) In fact "all fuels" excludes electricity in Balestra (1967).

$$J_t^{g*} \equiv J_t^g - (1 - \phi^g)J_{t-1}^g \tag{2.8.8}$$

and the new demand for all fuels

$$J_t^* \equiv J_t - (1 - \phi)J_{t-1} \tag{2.8.9}$$

where J_t is the quantity demanded of all fuels in British Thermal Units. This new demand for gas is assumed to depend only on the relative price of gas and the total new demand for all fuels:

$$J_t^{g*} = \alpha_1' + \alpha_2' J_t^* + \alpha_3' (P^g/\pi)_t \tag{2.8.10}$$

(This is the "basic equation" (p. 59). A form including the relative price of all other fuels was also tried.) Hence, adding a disturbance term,

$$J_t^g = \alpha_1 + \alpha_2 J_{t-1}^g + \alpha_3(P^g/\pi)_t + \alpha_4\Delta'(I/\pi) + \alpha_5(I/\pi)_{t-1} \\ + \alpha_6\Delta'N_t + \alpha_7 N_{t-1} + \varepsilon \tag{2.8.11}$$

Estimation of this form presents the same problems as encountered by Houthakker & Taylor: using Ordinary Least Squares, the restriction on the parameters $\alpha_5\alpha_6 = \alpha_4\alpha_7$ will not in general be satisfied; and biased estimators will be obtained because of the inclusion of a lagged endogenous variable. In fact Balestra's results imply a negative scrapping rate ϕ^g. He showed that the problem may be overcome in two ways: by using a different method of estimation based on instrumental variables which yields consistent estimators —this gave a plausible estimate of ϕ^g; or by constraining ϕ^g to be zero and using $\Delta'J^g$ as dependent variable—this he argued is reasonable for an appliance stock of low average age since the scrapping rate will then be very low.

The model was estimated firstly from cross-section data for 1962 and then from pooled cross-section and time-series data. This demonstrated that pooling yields improved estimates if additional variables are included to explain cross-section variations—Balestra found it necessary to allow for variations in weather between states, and for variations in the availability of a gas supply.

To investigate the hypothesis that the share of gas in replacement demand was different from its share of new demand, Balestra estimated

$$J_t^g = \alpha_1' + \alpha_3'(P^g/\pi) + \alpha_4'\Delta'J_t + \alpha_5'\phi J_{t-1} + (1 - \phi^g)J_{t-1}^g + \varepsilon \tag{2.8.12}$$

which is equivalent to (2.8.10), apart from the error term, if $\alpha_4' = \alpha_5'$.

2.9 The total demand for fuel

The use of the total demand for fuel as an explanatory variable implies a fundamentally different approach to the theory of demand, which we shall call the "Market Shares" approach, from that used in Chapter 1, the "classical" approach. It has also been adopted in the Ministry of Power's model of the whole British fuel economy described in Forster & Whitting (1968). The essential characteristic of a Market Shares model is that it starts by treating the market for an individual fuel as part of a larger market, for all fuels: ". . . the demand for gas is a derived demand, derived from the demand for total fuels" (Balestra (1967) p. 30). Typically the model takes the following form: the quantity variables are adjusted for variations in natural conditions and converted to a common unit. The total quantity demanded of all fuels is expressed as a function of real income and household wants variables. The market share of any fuel is assumed to depend only on relative fuel prices. Therefore demand for it is a function of all the fuel prices, the total demand for all fuels, and a random disturbance. For consistent estimation it is necessary that the structure be recursive so that the total demand for all fuels is a predetermined variable in the equations for individual fuels, independent of the error terms.

This Market Shares model suffers, however, from two weaknesses.[6]

(1) A severe restriction is imposed on the cross-elasticities: the coefficients of any fuel price in all the fuel equations must sum to zero. There is evidence that realistic restrictions would lead to improved estimates of cross-elasticities: whilst single equation models without restrictions have in some instances yielded plausible estimates, as in Houthakker (1951), "Past experience has shown that the tendency to select related prices for inclusion in a single equation model on grounds of improved explanation has often led to the adoption of very strong substitution relationships. These have often been greater in importance than the own-price of the commodity in question and the more general implications of these results are somewhat disturbing" (Brown (1958) p. 17).

But it seems unrealistic to suppose that the total demand for fuel, whatever the units of measurement, would remain constant if the general level of fuel prices rose relative to other prices. Even a change in the price of one fuel is likely to change the total demand. For example, a fall in the price of electricity would tend to increase the demand associated with some appliance types without there being an opposite substitution effect on the quantities of other fuels demanded, because there is no close substitute amongst other-fuel appliances. Income and tastes are likely to affect market shares, because of the variety in the types and quality of appliances associated with any fuel.

In particular applications, the implied restrictions have been even more severe and the resulting cross-elasticities implausible. In Balestra (1967), the basic model—equation (2.8.10)—excluded all prices except the own price P^g; thus symmetry and consistency with the assumption about sums of coefficients are incompatible. When the relative price of all other fuels was included, the coefficient was negative.

(6) See Baxter & Rees (1968) pp. 277-278.

(2) Let us suppose that within the particular time-period and area under study, the above restriction is a reasonable approximation to reality. Then the approach runs into difficulties with the choice of a common unit of measurement. For example, consider the use of thermal units as in Balestra (1967). It would seem most reasonable to suppose that, in the space-heating market, the demand for effective heat would be the total which remained constant, given wants and income. But the coefficient for converting the quantity demanded of electricity in kilowatt-hours into effective heat units depends on the composition of the appliance stock and the efficiency of appliances, data on which are lacking. For other uses, e.g. lighting, the calculation would be further complicated by the inappropriateness of heat units. (There is also the problem that the efficiency of appliances may change over time, but this applies to the classical model as well.) The use of the more usual crude aggregate conversion factors is likely to lead to large errors in weighting. This will give an erroneous total fuel series and hence misleading estimates of all the parameters.

2.10 Wigley (1968)[7]

It is possible to impose a restriction on the cross-elasticities without introducing the total demand for fuels or excluding income and household wants from the individual fuel equations, as in Wigley (1968), where the following equation is used:

$$\log (V/N) = \beta_1 + \beta_3 \log W + \beta_7 \log (E/\pi) + \beta_8 \log (P^e/p^f) + \varepsilon \quad (2.10.1)$$

where V is the quantity of electricity consumed, valued at 1958 prices, in pounds,

N the population,

W the ratio of the average annual temperature to the long-run average,

E total consumption expenditure per head,

π an all-commodities price index,

P^e the average price of electricity,

p^f an index of the prices of all fuels, including electricity, using expenditure per head at 1958 prices as weights.

Equations of the same form were estimated for three other fuels—solid fuel, oil other than motor spirit, and gas. The imposition of the restriction does not depend on the simultaneous estimation of these equations and in fact they were estimated independently by Ordinary Least Squares. But data are required on the prices and consumption of the other fuels.

The form of the model implies that there are no substitution effects if all fuel prices, or all other prices, change, but it does allow for differences in

(7) pp. 60-67.

the effect of income and tastes on different fuels. The total expenditure coefficients obtained support the hypothesis that these effects do differ and therefore that the restriction imposed by the Market Shares model is unrealistic.

The equations were estimated from annual time-series for the United Kingdom 1955-65; giving the results in Table 2.10.1.

TABLE 2.10.1. *Wigley (1968) results*
Source: Wigley (1968) pp. 62, 64

Parameter	β_3	β_7	β_8	
Estimate	−1·92	3·65	−0·81	$R^2 = 0.99$
$|t|$	4·0	14·8	2·7	

	Elasticity of electricity demand with respect to the price of			
	Solid fuel	oil	gas	electricity
1955	0·407	0·0175	0·189	−0·614
1960	0·309	0·0394	0·165	−0·513
1965	0·209	0·0564	0·193	−0·459

As noted by Wigley, the rather high total expenditure elasticity of 3·65 probably reflects trends in omitted taste variables—and the effect of past values in producing an upward trend in appliance ownership. It is less obvious in this model what the effect of using the average price of electricity is, as the other components of the index of fuel prices are also average prices. Calculations on the figures tabulated by Wigley show that P^e/p^f tends to fall when E/π rises, suggesting that, as in the Stone (1954) model, the effect is to bias downward both the expenditure and own-price coefficients.

2.11 Fisher & Kaysen (1962)

Fisher & Kaysen developed both a short-run and a long-run model in their analyses of the household demand for electricity in the United States. Their short-run model is of interest because it is developed from an expression in which the appliance stock is disaggregated. Their derivation of a form in which only the aggregate stock appears therefore lays bare the assumptions behind the model in Houthakker (1951), and the short-run models in Doumenis (1965) and Balestra (1967).

They started from the identity

$$Q_t \equiv \sum_a Z_t^a U_t^a \qquad (2.11.1)$$

where the summation is over appliance types,
Q is the total quantity demanded of electricity

Z^a the number of appliances of type a, times their mean wattage rating, times the proportion of this rating in use during an hour of average use.

and hence U^a is utilization in hours. Utilization is assumed to be a function of the form

$$\ln U_t^a = \beta_{1,a} + \beta_{7,a} \ln (I/\pi)_t + \beta_{8,a} \ln (P^e/\pi)_t \qquad (2.11.2)$$

where I is income per head,

π an all-commodities price index,

P^e the average price of electricity.

Their equation for Q_t is

$$\ln Q_t = \ln Z_t + \beta_1 + \beta_7 \ln (I/\pi)_t + \beta_8 \ln (P^e/\pi)_t + \varepsilon_t \qquad (2.11.3)$$

where $$Z_t = \sum_a Z_t^a,$$

which implies that (1) $\beta_{j,a} = \beta_j, j = 1, 7, 8$; that is, the elasticities and utilization in hours are the same for all types of appliances; and (2) the composition of the appliance stock remains unchanged.

They show that, if these conditions do not apply, it is still satisfactory to use this form if the composition of the stock changes only slowly and at a steady rate. In these circumstances, β_{14} introduced in (2.11.5) below includes the effects of this change on average utilization, and the price and income elasticities are averages which vary between areas with the consumption of the appliance stock.

As Fisher & Kaysen did not have the required data on appliance ownership, they made the same assumption as Balestra:

$$Z_t = (1 + \kappa) Z_{t-1} \qquad (2.11.4)$$

Thus, taking first differences,

$$\Delta \ln Q = \beta_{14} + \beta_7 \ln (I/\pi) + \beta_8 \Delta \ln (P^e/\pi) + v \qquad (2.11.5)$$

where $$\beta_{14} = \ln (1 + \kappa).$$

This was estimated separately for each state from annual time-series 1946-57, and subsequently for groups of states classified by degree of urbanization. The results were poor for individual states but supported the hypothesis that the price and income elasticities (which may be interpreted here as short-run ones) for different appliances differed widely and cross-section differences in income and tastes—as correlated with urbanization—were associated with differences in the composition of the appliance stock such that the average elasticities varied regionally.

The long-run model consisted of a single equation in which the appliance stock was the dependent variable. In conclusion it was stated (p. 117) that the elasticity of demand with respect to price or income is the sum of two elasticities—the short-run one from (2.11.5) plus the one from the long-run appliance demand equation. Thus the assumption is made, contrary to that in Cramer (1959), that there will be no effect in the long run on average utilization associated with changes in the appliance stock.

2.12 Conclusions

Some conclusions may now be drawn about which of the modifications listed in Chapter 1 should be incorporated in our model given the data available.

We concluded earlier, in 1.8, that the data available limit our choice of model to two sorts: either one which includes a very simple equation for electricity and similar equations for other fuels and appliances; or one which comprises a single equation for electricity of a relatively elaborate form. The first, simultaneous equation alternative is attractive in that it allows investigation of the possible improvements to estimates from imposing the theoretical restrictions which involve the coefficients of more than one equation. Wigley, however, has already used the available aggregate British statistics to estimate a model containing four fuel demand equations; and our additional data is limited to variables appearing in the electricity equation. On the other hand these data do permit estimation of a single equation model which incorporates more of the modifications listed in 1.7 than those models previously estimated. This promises to be fruitful in several respects: there is a need to analyse the effects on utilization of changes in prices and income in the long run, that is when appliance ownership also changes; Houthakker's results suggest that an improvement in explanatory power may be achieved by including hours of daylight as well as temperature in the demand equation—and allowing for a non-linear response—and by including lagged prices and income to allow for habits in utilization; Fisher & Kaysen's results suggest that a considerable improvement can be achieved by disaggregating by appliance type.

We shall therefore develop a single equation model including lagged values, temperature, and hours of daylight as explanatory variables, with disaggregation by appliance type and with a form determined by the complementarity condition. It will be used to analyse the effects of natural and economic variables on utilization, taking appliance ownership as given and predetermined. The properties and interpretation of the estimates obtained will be examined in the following four chapters.

CHAPTER 3

THE MODEL

3.1 Introduction

In this chapter the general form of the demand function is derived.

Assumptions are introduced in 3.3-3.5 which lead to a simple interpretation of the demand parameters. The effect of the invalidity of these assumptions on the interpretation is examined in 3.6 and, as a result, two modifications to the model are introduced, in 3.7 and 3.8.

The term "appliance" will be used throughout to mean any equipment which uses fuel, including for example light bulbs. For simplicity of exposition it is assumed that all appliances are "fuel-specific", that is, use only one fuel to fulfil their main function, so that consumers have to change from one appliance to another in order to use a different fuel. In the period under study the number of appliances in use which were not fuel-specific was negligible.

Where there is a difference between the number of households *having* a particular type of appliance and the number *owning* it, the number having is the relevant variable for the present analysis. We shall, however, follow the standard practice of referring to it as an ownership variable.

3.2 The basic identity

Let the quantity demanded by an individual consumer, h, due to use of appliances of type a, on day d of month m, be Q_{dm}^{*ah} kilowatt-hours. Then we may write

$$Q_{dm}^{*ah} \equiv (Alur)_{dm}^{ah} \qquad (3.2.1)$$

where $A = 0$ if no appliance of type a is owned,
    $= 1$ otherwise,
  l is the (maximum) wattage rating of all the appliances owned of type a, in kilowatts,
  u the time, in hours, during which the consumer demands the services of the appliances, or the appliances are on to meet a later demand,

    r the mean proportion of the wattage rating in use during this time $(0 < r \leqslant 1)$.

Subscripts and superscripts appearing outside brackets refer to each of the terms inside. For the rest of this chapter the time-subscripts will be suppressed.

The number of hours u may be greater than the number of hours during which the appliances are using electricity as some appliances meet a continuous demand without consuming electricity continuously.

The total quantity demanded on unrestricted tariffs by consumer h is the sum of expressions (3.2.1) over all types of appliance, a, other than off-peak ones:

$$Q^{*uh} \equiv \sum_a (Alur)^{ah} \qquad (3.2.2)$$

3.3 The demand equation for an individual consumer

In general A, l, u, and r are each a function of income, prices, and other variables; the four functions for each type of appliance are interdependent one with another and with the functions for other types.

Four assumptions are made about the properties of these functions:

Assumption 3.3.1

Appliances may be classified such that

(i) within each class, a, the parameters in the functions A, l, u, and r for each type of appliance and the interdependences between types are such that, in a set of four aggregate functions for the class as a whole, the parameters are independent of the composition of the stock of appliances within the class;

(ii) the aggregate functions for different classes are independent.

It is further assumed that these two conditions are satisfied by classifying appliances into four categories—space-heaters, water-heaters, cookers, and lighting and sundry appliances—and separating out off-peak space-heaters and water-heaters.[1]

Assumption 3.3.2

Either (i) let $Z^{ah} \equiv (Al)^{ah}$

           and $U^{ah} \equiv (ur)^{ah}$ (3.3.1)

then Z^{ah} and U^{ah} are independent functions in the sense that the parameters of U^{ah} measure the effect of its arguments on Q^{*ah} given Z^{ah} constant, and the parameters of Z^{ah} the effect of its arguments on Q^{*ah} given U^{ah} constant;

(1) For definitions of the classes, see Chapter 4.

or (ii) let $Z^{ah} \equiv A^{ah}$
   and $U^{ah} \equiv (lur)^{ah}$ (3.3.2)

then Z^{ah} and U^{ah} are independent functions in the same sense.

Alternative (i) implies that a consumer increases the installed load in class a, l^{ah}, only if he has a greater demand for the services of appliances in this class; (ii) that a consumer increases l^{ah} only to satisfy the same want in a shorter time, or in the same time more conveniently.

The total effect of a change in a variable hence consists of two parts: its effect on the appliance stock—which we shall call the Z *effect*; and its effect on the utilization of a given appliance stock—which we shall call the U *effect*.

The basic identity (3.2.2) may be rewritten as

$$Q^{*ah} \equiv \sum_{a=1}^{4} (ZU)^{ah}$$ (3.3.3)

Utilization will hereafter mean U rather than u unless stated otherwise.

Assumption 3.3.3

The utilization of appliances in class a may be expressed as a linear function of N variables X_j^h (the value of N and the list of variable numbers j to be specified later) to be explicitly included and a large number of other variables the net influence of which may be represented by a random error term, ε^{ah}:

$$U^{ah} = \sum_j \beta_{j,a}^h X_j^h + \varepsilon^{ah}$$ (3.3.4)

Assumption 3.3.4

Each appliance stock Z^{ah} is predetermined in the demand equation for electricity: the demand for each appliance is a function of variables exogenous to the model and independent of the error terms in the utilization functions.

It is also assumed that the difference between actual demand, Q^{*uh}, and recorded demand, Q^{uh} say, may be represented by a random error term ε^{*h}. Hence the recorded demand for electricity by consumer h is given by

$$Q^{uh} = \sum_{a=1}^{4} Z^{ah} \left(\sum_j \beta_{j,a}^h X_j^h + \varepsilon^{ah} \right) + \varepsilon^{*h}$$ (3.3.5)

3.4 Aggregation over consumers

On aggregating the functions for all consumers in each Electricity Board Area, two further assumptions are made.

Let the number of consumers in any given Area be C.

Assumption 3.4.1

Let

$$\beta_{j,a}^* = \sum_h \beta_{j,a}^h / C$$

Then for all j, a,

$$\beta_{j,a}^* = \frac{\sum_h \beta_{j,a}^h Z^{ah} X_j^h}{\sum_h Z^{ah} X_j^h} = \frac{\sum_h \beta_{j,a}^h X_j^h}{\sum_h X_j^h} \tag{3.4.1}$$

and $\beta_{j,a}^*$ is the same in all Areas and time-periods.

Not all electricity consumers in any given Area had available a supply of any given other fuel, that is, were able to obtain it without incurring a non-negligible delivery or connection charge. Of those that did have a supply available, not all had an appliance specific to that fuel in any particular appliance class. Because of the broadness of these classes, there may have been some owners of other-fuel appliances who did not substitute use of these for use of their electrical appliances within the range of prices experienced because they were not sufficiently close substitutes. Only for those who had a supply of another fuel and an appliance specific to that fuel which they regarded as a substitute for their electrical appliances would a change in the price of the other fuel have had a non-zero substitution effect on U^{ah}. Thus this assumption implies that for each other fuel and each class of appliance, these consumers were the same proportion of the total number of consumers in each Area and time-period.

Assumption 3.4.2

$$\frac{\sum_h Z^{ah} X_j^h}{\sum_h Z^{ah}} = \frac{\sum_h X_j^h}{C} \quad \text{all } j \tag{3.4.2}$$

Hence on summing over h from 1 to C, dividing by C, and writing

$$Q^u \equiv \sum_h Q^{uh} \tag{3.4.3}$$

$$X_j^* \equiv \frac{\sum_h X_j^h}{C} \tag{3.4.4}$$

$$Z^a \equiv \frac{\sum_h Z^{ah}}{C} \tag{3.4.5}$$

$$\varepsilon \equiv \frac{\sum_a \sum_h Z^{ah} \varepsilon^{ah} + \sum_h \varepsilon^{*h}}{C} \tag{3.4.6}$$

we obtain

$$\frac{Q^u}{C} = \sum_a Z^a \sum_j \beta_{j,a}^* X_j^* + \varepsilon \qquad (3.4.7)$$

Defining mean utilization as

$$U^a \equiv \frac{\sum_h U^{ah}}{C} \qquad (3.4.8)$$

and writing

$$\varepsilon^a \equiv \frac{\sum_h \varepsilon^{ah}}{C} \qquad (3.4.9)$$

(3.4.7) may be written as

$$\frac{Q^u}{C} = \sum_a Z^a (U^a - \varepsilon^a) + \varepsilon \qquad (3.4.10)$$

3.5 Variables in the utilization functions

The set of determining variables, X_j^h, may be specified on the basis of our conclusions in Chapters 1 and 2. Not all the variables should appear in each function U^a; restrictions on the parameters will be introduced in Chapter 6.

The set includes a constant—

$$X_2^h \equiv 1, \qquad (3.5.1)$$

some function of temperature, W, and hours of daylight, L—

$$X_3^h \equiv X_3(W^h) \qquad (3.5.2)$$

$$X_4^h \equiv X_4(L^h) \qquad (3.5.3)$$

and functions of the current and lagged values of the mean total consumption expenditure per household (E), the prices of electricity (P^e) and its main substitutes—gas (P^g) and solid fuel (P^c)—and all other prices, represented by an index (π).[2]

Given the low substitutability of fuels for other commodities and the limited scope for substitution between fuels open to most households given their stock of appliances, it is likely that the rate of increase of U^a rapidly approaches zero as P^e falls or E, P^c or P^g rises, for classes of appliances for which these variables have non-zero parameters in U^a. A convenient way to specify this is to make U^a linear in the logarithms of the variables. We therefore define variables as follows:

$$X_7^h \equiv \text{lag ln } (\bar{E}/\pi)^h \qquad (3.5.4)$$

$$X_8^h \equiv \text{lag ln } (P^e/\pi)^h \qquad (3.5.5)$$

$$X_9^h \equiv \text{lag ln } (\bar{P}^c/\pi)^h \qquad (3.5.6)$$

$$X_{10}^h \equiv \text{lag ln } (\bar{P}^g/\pi)^h \qquad (3.5.7)$$

(2) See Chapter 4 for the full definitions of these variables, and our reasons for using total consumption expenditure rather than income.

where lag is an operator denoting that the variable is a weighted average of current and lagged values of the operand, and bars denote seasonally adjusted variables.[3]

The expressions (3.5.2)-(3.5.7) may be written in the general form

$$X_j^h \equiv X_j^* * (X_j^{*h}) \tag{3.5.8}$$

Let X_j be defined as follows:

$$X_j \equiv X_j^* * \left(\frac{\sum X_j^{*h}}{C}\right) \tag{3.5.9}$$

The data permit calculation of X_j but not of the means X_j^* given by (3.4.4). In order to replace the demand equation (3.4.7) by an expression which does not contain the X_j^*, the following assumption is made:

Assumption 3.5.1

X_j and X_j^* differ only by a constant, for all j.

Hence (3.4.7) may be replaced by

$$\frac{Q^u}{C} = \sum_a Z^a \sum_j \beta_{i,a} X_j + \varepsilon \tag{3.5.10}$$

where $\qquad\qquad\qquad\qquad X_2 \equiv 1,$

$$\beta_{j,a} = \beta_{j,a}^* \qquad \text{all } a, \text{ all } j \neq 2$$

The variables in (3.5.2)-(3.5.7) will be called the *exogenous* variables. The variables Z^a will be called the *appliance ownership*, or *predetermined*, variables, A^a the *proportion of consumers owning* appliances of type a, and $(Al)^a$ the *installed load per consumer*.

Finally we make

Assumption 3.5.2

Given the previous assumptions, the utilization of each class of appliances is correctly specified as a function only of those exogenous variables in (3.5.2)-(3.5.7); in particular, it is not a function of any variables representing household wants.

Given these assumptions, each coefficient may be simply interpreted as a measure of the U effect of the corresponding exogenous variable.

---

(3) For the determination of the weights and the justification for using seasonally adjusted figures, see 4.15 and 4.19.

3.6 The validity of the assumptions

The validity of assumption 3.3.1

Attempting to design a classification of appliances for which assumption 3.3.1 is valid, we are in the usual dilemma: with too few classes, the homogeneity condition (i) will not be met; with too many, the independence condition (ii) will not be met. Four classes have been used because any more would present data and estimation problems; any fewer would mean an unacceptable degree of aggregation. As a result, neither condition is exactly fulfilled. Firstly, the parameters for different appliances within classes are likely to be different. This is not too serious as long as the composition of the stock within a class does not vary greatly. If it does vary, there will be an effect on the mean utilization of that class. For example, if there is a shift towards owning more efficient appliances as total real consumption expenditure rises, mean utilization will tend to fall. Secondly there is some interdependence between the four classes, changes in the stock of appliances in one class having an effect on the utilization of other classes. For example, the utilization of cookers depends on the ownership of kettles, which are included in the lighting and sundry class.

As the invalidity of other assumptions also gives rise to effects on utilization which are not U effects, discussion of the problem of how to allow for them is postponed until the next section. It is convenient to have a name for them. We shall call any effect on the utilization of a given class of appliance associated with a change in the level or composition of the stock of appliances in the same or a different class a *Z-U effect*. The total effect of a change in a variable is then in general the sum of three parts—the Z, U, and Z-U effects.

The validity of assumption 3.3.2

The alternative assumptions 3.3.2 (i) and (ii) giving rise to independent appliance ownership and utilization functions represent extreme situations. Neither is exactly valid for any of the four classes used. Hence for each class of appliance a, changes in the ownership level Z^a have Z-U effects on utilization U^a.

The validity of assumption 3.3.3

Assumption 3.3.3 is made because use of any non-linear function would yield an intractable demand equation on summation over appliance types. On the arguments advanced in 3.5 above, the assumption is consistent with a plausible specification of the equation. Its restrictiveness will be reassessed after estimation, by examination of the pattern of residuals.

The validity of assumption 3.3.4

The true appliance ownership levels, Z^a, should be very nearly independent of the error term in the demand equation for electricity, ε: the size of this error in a given period may depend on variables which affect the net acquisition of appliances in the same period but this is a small part of the total stock for all of the classes used; the error is unlikely to depend on variables which affected net acquisitions in earlier periods.

The presence of measurement error due to the use of sample data and interpolated values would produce negative correlation between the estimated values of Z^a used and ε, and hence biased estimators. The errors due to interpolation should not however be large, as indicated in 1.4 above. The likely size of sampling errors is examined in 4.11 below.

The validity of assumption 3.4.1

It is likely that there were variations in the proportions of consumers with substitute gas and solid-fuel appliances (and hence in the coefficients $\beta^*_{j,a}$) due to variations in relative fuel and appliance prices used and in the availability of the fuels.

For solid fuel there is no information with which to improve the specification. But for gas the data are sufficient to enable us to allow for the effect of variations in the availability of a supply; a modified aggregate function will be derived in 3.8 below.

The validity of assumption 3.4.2

If the exogenous variables X^h_j and appliance ownership variables Z^{ah} are correlated within the groups of consumers in each Electricity Board Area, then the calculated unweighted mean X_j will differ from the required mean weighted by Z^{ah}, resulting in biased estimators. The data are insufficient to permit calculation of the weighted means.

For temperature and illumination, the weighted and unweighted means should differ very little. The calculated variables measure only differences due to natural conditions and not those due to the type of dwelling occupied. If variations in natural conditions within Areas affect appliance ownership then inter-Area differences in the means will be subject to error, but this should be slight given the number of other determinants of ownership. Within Areas the correlation between changes in natural conditions experienced by different consumers is very high so that errors in the measurement of intertemporal differences should be negligible.

The past and present values of prices and total consumption expenditure included as arguments of the utilization function U^a may be expected also to be determinants of Z^a. Hence for any variable which does not take the same value for all consumers within a particular Area, there is an error involved in using the unweighted mean. Determination of the consequences is left until Chapter 4 when the net effect of all the errors in the calculated series can be assessed.

The validity of assumption 3.5.1

For none of the price and natural variables were the variations in the shape of the distributions within Areas large enough to produce more than small variations in $X_j^* - X_j$. For $j=8$ the difference was always zero since the price of electricity P^e as defined[4] was constant within each Area. It has been shown[5] that the difference $X_j^* - X_j$ is constant if X_j^h is lognormally distributed with a constant variance, as may be approximately true for real total consumption expenditure, E/π.

The validity of assumption 3.5.2

The utilization of appliances by a household depends to a considerable extent on variables which have been omitted, such as the size and composition of the household, the amount of heat insulation, the number and size of rooms and the number of meals eaten at home. For many of these variables data are nonexistent; for the remainder they are limited to unweighted means for all consumers in each region. Typically these means show only a gradual change over time and inter-Area variations within the margins of error of the data.

For a variable which is uncorrelated with appliance ownership, the unweighted means would be the correct values to use. But given their properties as just indicated, it is unlikely that reliable coefficients could be obtained. The exclusion of such variables will probably lead to bias in the coefficients of included variables which changed gradually, but we would not expect this to be large because the variables in this category are relatively few and the changes in their mean values should be small.

Some omitted variables, for example, the amount of pipe lagging, may be correlated with appliance ownership although not determinants of it, because both depend on some third variable, typically income. For these variables also, it is unlikely that reliable coefficients could be obtained even with accurate data on the weighted mean values, because of their high correlation with other included variables. Their omission will tend to produce bias in coefficients, particularly of the total expenditure variable, though no variable stands out as likely to cause serious bias.

The most important consequences arise from the omission of variables

---

(4) See 4.14.

(5) See Aitchison & Brown (1957) p. 122. On the same aggregation problem, see also Prais & Houthakker (1955), Farrell (1954).

which are correlated with ownership because they are determinants of it. The coefficient of such a variable in a correctly specified utilization function U^a would in general measure two effects: firstly, the U effect of a change in the value of the variable for existing owners of a given stock of appliances. If the omitted variable is correlated over time and between Areas with some included exogenous variable then the coefficient of this variable will take up the explanation of this U effect. For example, if the scale effect of household size dominates the income effect for utilization of electrical appliances,[6] and household size is negatively correlated with real total household consumption expenditure, E/π, over time and between Areas, then the coefficient of E/π will measure the sum of the positive U effect of an increase in E/π and the negative U effect of a fall in household size. Since some variables of this type are not correlated with any included variable between Areas, some U effects will go unexplained in the error term.

Secondly, the coefficient would measure the Z-U effect of a change in the weighted mean value of the variable associated with a change in appliance ownership Z^a brought about by a change in another or the same variable, and arising because the marginal utilization of new owners, and of existing owners changing their installed load, is different from the mean. If the change in Z^a was caused by a variable included in the U^a function, its coefficient will take up the explanation of the Z-U effect; if by some other variable, then either the coefficient of a correlated variable will take up the explanation or it will go unexplained in the error term.

3.7 Modification to allow for the Z-U effects

Since there were large variations in appliance ownership over time and between Areas, it is likely that the Z-U effects were large. As it is impracticable to include all the variables which explain them, some modification of the model is desirable, so that as far as possible the coefficients of the variables included measure only the effects of changes in these variables, and the error variance is reduced.

The modification adopted is the inclusion of the predetermined ownership variables in the utilization functions as proxies for the omitted variables. In each function U^a the reciprocal of the corresponding ownership variable is included:

$$X_1^{(a)} = \frac{1}{Z^a} \qquad (Z^a > 0) \tag{3.7.1}$$

This form is used firstly because it is plausible that successive unit increments in Z^a were associated with successively smaller changes in   ean

(6) These effects are defined in 4.17.

utilization—and secondly because the equation to be estimated then has a simple form. Augmenting (3.5.10),

$$\frac{Q^u}{C} = \beta_1 + \sum_a Z^a \sum_{j>1} \beta_{j,a} X_j + \varepsilon \qquad (3.7.2)$$

where
$$\beta_1 \equiv \sum_a \beta_{1,a}$$

$\beta_{1,a}$ measures the average Z-U effect of changes in Z^a.

The rapid rise in the ownership of off-peak appliances was probably associated with Z-U effects that were large (absolutely) for the size of the change in Z^a. As the necessary data are available, two more variables are included to measure these effects:

$X_5 \equiv Z^{o1}$, the ownership of off-peak space-heaters;
$X_6 \equiv Z^{o2}$, the ownership of off-peak water-heaters.[7]

With these changes in the model, the coefficients of the prices and total consumption expenditure may be interpreted as follows: $\beta_{j,a} (j=7, \ldots, 10)$ measures the sum of the U effect of changes in the present and past values of the exogenous variable of which X_j is a function and the Z-U effect of changes in these values, less the average Z-U effect on U^a measured by $\beta_{1,a}$.

We would expect U effects to be small whilst Z-U effects may be large and opposite in sign. Hence it is quite possible for $\beta_{j,a}$ to have the opposite sign to the U effect—for example, for the coefficient of the price of electricity to be positive.

3.8 Modifications to allow for variations in the availability of a gas supply

A further modification to the model follows from the relaxation of assumption 3.4.1 referred to in 3.6 above to take account of variations in the availability of a gas supply.

Of those consumers who would incur a positive cost in obtaining a gas supply at a particular point in their house, not all would incur the same cost: whereas some might live distant from a gas main, others would already have gas pipes in the house. The data available permit consumers to be divided reasonably satisfactorily into two classes—those who would regard themselves as having a supply available at negligible cost and those who would not.[8] Assumption 3.4.1 may therefore be replaced by

(7) For definitions see 4.8 and Chapter 7. The number of off-peak appliances in the
    other classes was negligible.
(8) For the definitions, see 4.12.

Assumption 3.8.1

For all variables j and appliance classes a, the distribution of values $\beta^h_{j,a}$ for households with a gas supply available has the same mean $\beta^G_{j,a}$ in all Areas and time-periods, and the values $\beta^h_{j,a}$ for households without a supply the same mean $\beta^{NG}_{j,a}$; within both groups, $\beta^h_{j,a}$ is uncorrelated with $Z^{ah}X^h_j$ and X^h_j; the coefficient of the price of gas $\beta^h_{10,a}=0$ for all households without a supply.

Let G denote the proportion of consumers with a supply available, superscript G variables relating to these consumers, and superscript NG variables relating to the other consumers. The mean demand of all consumers may then be written:

$$\frac{Q^u}{C}=\sum_a Z^a\left\{\left[\beta^G_1 G\left(\frac{Z^{aG}}{Z^a}\right)X^{(a)G}_1+\beta^{NG}_1(1-G)\left(\frac{Z^{aNG}}{Z^a}\right)X^{(a)NG}_1\right]\right.$$

$$+\left(\frac{Z^{aNG}}{Z^a}\right)\beta^{NG}_{2,a}+\sum_{j=3}^9\left[\beta^{Ga}_{j,}G\left(\frac{Z^{aG}}{Z^a}\right)X^G_j+\beta^{NG}_{j,a}(1-G)\left(\frac{Z^{aNG}}{Z^a}\right)X^{NG}_j\right]$$

$$\left.+\beta^G_{10,a}\left(\frac{Z^{aG}}{Z^a}\right)G\text{ lag ln }(\bar{P}^g/\pi)+\left[\left(\frac{Z^{aG}}{Z^a}\right)\beta^G_{2,a}-\left(\frac{Z^{aNG}}{Z^a}\right)\beta^{NG}_{2,a}\right]G\right\}$$

$$+\varepsilon \tag{3.8.1}$$

Information is lacking on Z^{aG} and Z^{aNG}. The modification made is to add to each function U^a specified in 3.5 above one term $\beta_{11,a}G$ and to replace $\beta_{10,a}$ lag ln $(\bar{P}^g/\bar{\pi})$ by $\beta_{10,a}G$ lag ln $(\bar{P}^g/\bar{\pi})$.

An assumption which leads to the elimination of the unknown quantities from (3.8.1) is the following:

Assumption 3.8.2

Either (i) $Z^{aG}=0$; $X^G_j=X^{NG}_j$, $j=3, ..., 9$;[9]
or   (ii) $Z^{aNG}=Z^{aG}$; $\beta^G_j=\beta^{NG}_j$, $j=1, 3, 4, ..., 9$.

Writing $\beta_1\equiv\beta^{NG}_1$

$\beta_{j,a}\equiv\beta^{NG}_{j,a}$, $j=2, ..., 9$

$\beta_{10,a}\equiv\beta^G_{10,a}$

$\beta_{11,a}\equiv\beta^G_{2,a}-\beta^{NG}_{2,a}$

if (i) applies for all a, then

[9] For $Z^{aG}=0$, it is necessary to define $X^{(a)G}_1=0$ to avoid the implication of positive demand with a zero stock. It is then also necessary to assume $\beta_1^{NG}=0$ to get (3.8.2).

$$\frac{Q^u}{C}=\beta_1+\sum_a Z^a\left\{\beta_{2,a}+\sum_{j=3}^{9}\beta_{j,a}X_j\right\}+\varepsilon \tag{3.8.2}$$

if (ii) applies for all a, then

$$\frac{Q^u}{C}=\beta_1+\sum_a Z^a\left\{\beta_{2,a}+\sum_{j=3}^{9}\beta_{j,a}X_j+\beta_{10,a}G\text{ lag ln }(\bar{P}^g/\bar{\pi})+\beta_{11,a}G\right\}+\varepsilon \tag{3.8.3}$$

Alternative (i) implies that gas appliances are regarded as substitutes for electrical ones and are preferred whenever gas is available.

Alternative (ii) implies that a stock of gas appliances is not a substitute for a stock of electrical appliances and, since the coefficient of the price of electricity does not vary with gas availability, that utilization of a gas appliance is not a substitute for utilization of an electrical one. Ordinarily this would mean that the U effect of a change in the price of gas was zero. Hence given assumption 3.8.2 and all the preceding assumptions (except 3.4.1),

$$\beta_{10,a}=\beta_{11,a}=0 \tag{3.8.4}$$

But the alternatives (i) and (ii) represent extreme situations and are unlikely to be valid for any of the four classes of appliances. Their invalidity means that the estimates of $\beta_{10,a}$ and $\beta_{11,a}$ are averages of functions of Z^{aG} and Z^{aNG}. Some empirical evidence on the probable range of these variables will be examined in 4.13. The invalidity of assumptions 3.4.2 on the absence of correlation between the appliance ownership and exogenous variables and 3.5.2 on the correct specification of the utilization functions means that the coefficients measure both a U effect and the Z-U effects of differences in the ownership of gas and electrical appliances between the two groups of consumers.

3.9 The final form

As a result of the modifications introduced in the previous two sections, expression (3.5.10) is replaced by

$$\frac{Q^u}{C}=\beta_1+\sum_a Z^a\sum_{j=2}^{11}\beta_{j,a}X_j+\varepsilon \tag{3.9.1}$$

where $$X_{11}\equiv G \tag{3.9.2}$$

and $$X_{10}\equiv G\text{ lag ln }(\bar{P}^g/\bar{\pi}) \tag{3.9.3}$$

replacing (3.5.7).

This is the final form of the demand equation for day d of month m. Aggregation over time will be performed in 4.4.

3.10 The error term

If the errors in the individual consumer's demand functions (3.3.5) are assumed to satisfy the following conditions—

$$\left.\begin{array}{ll} E\{\varepsilon^{ah}\}=E\{\varepsilon^{*h}\}=0, & \\ E\{\varepsilon^{ah}\varepsilon^{ak}\}=0 & h\neq k \\ \qquad\qquad =(\omega^a)^2 & h=k, \\ E\{\varepsilon^{*h}\varepsilon^{*k}\}=0 & h\neq k \\ \qquad\qquad =(\omega^*)^2 & h=k \end{array}\right\} \text{ all } a, h-$$

then from (3.4.6) the error term ε in the final form is distributed with mean zero and variance

$$\sum_a (\omega^a)^2 \frac{\sum_h (Z^{ah})^2}{C^2} + \frac{(\omega^*)^2}{C} \tag{3.10.1}$$

But errors for consumers within any Area were probably not independent so that Areas with a larger number of consumers, C, did not necessarily have a lower variance.

Given that there were larger rises in the ownership variables Z^a than in C, the variance (3.10.1) would almost certainly have risen over time. But many of the series used improved in accuracy over the period so that, *ceteris paribus*, more rather than less weight should be attached to the later observations.

It is therefore appropriate to specify the variance as more nearly constant than in (3.10.1). As the estimation problems which arise because aggregation over time-periods of different lengths is necessary become much more tractable if ε is homoscedastic,[10] the following assumption is made:

Assumption 3.10.1

The error terms ε for each day and each Area are independently distributed with mean zero and constant variance, σ^2.

3.11 Identification

The identifiability of the parameters in (3.9.1) depends on the form of the fuel supply functions.

For gas in the short run, that is, when the tariff is fixed, the marginal price was typically a step function of the quantity supplied; our variable, P^g, is a weighted average of the marginal rates, the weights varying with average consumption.[11] This would lead to underidentification if ε were correlated

(10) See 4.4.
11) See 4.15.

with any of the terms in the demand function for gas, but this seems unlikely unless there is serious specification error in (3.9.1).

For electricity in the short run, the marginal price for any one consumer was also typically a step function of the quantity supplied. Consequently if, as in the variant model developed in Chapter 7, an average of the marginal rates is used, it is necessary to specify a function defining the tariff to overcome the identification problem.

In this model, the final rate, P^{ef}, is used as the price variable. It is not perfectly correlated with the average marginal price because the proportion of consumers having each rate as their marginal rate varied and other rates changed relative to the final rate. It is used because there are estimation problems associated with the variant model which preclude analysis of the effects of varying the specification of the model; and because the error involved should be a small one[12] the effect of which can be checked by using the variant model.

In the period under study, there were no quarters in which consumption was constrained by supply limitations for more than a few hours, so that the existence of an upper limit to the supply at the final rate can be ignored.

As to the long run, during 1955-68 each Electricity Board changed its final rate, P^{ef}, between three and seven times, and there were other changes in tariffs as well. Although the level of domestic demand would be one determinant of such changes:

(1) Changes in generating capacity are determined by changes in total demand, not just domestic demand. As non-domestic demand was subject to different rates of change, changes in the domestic tariff would tend to occur independently of changes in domestic demand.

(2) The relationship between current demand and current tariffs was weakened by the policy of making infrequent changes in domestic tariffs.

(3) The quantity variable used in this analysis is quarterly sales; the more important determinant of tariffs was the Simultaneous Maximum Demand from all consumers.[13] This changed at a different rate from quarterly sales.

(4) Given that the time-lag between the decision to build a power station and the addition to generating capacity is several years, the current tariff depended on forecasts made several years earlier of current and future demand. Thus to the extent that forecasts were inaccurate, there would only be a weak supply relationship between current actual consumption and current and past prices.

(12) See 4.14.
(13) See the details of the Bulk Supply Tariff given annually in the Electrical Times *Electricity Supply Handbook* and Meek (1968).

(5) There were variations in the final rate P^{ef} between Electricity Boards, mainly because of variations in the wholesale price at which electricity was supplied to them by the Central Electricity Generating Board. As these were determined with reference to fuel costs in each Area and each Board's contribution to the total national Simultaneous Maximum Demand from all users, the relationship to domestic quarterly sales on the supply side was a weak one.

It can therefore be concluded that little inaccuracy will result from regarding the parameters in the demand function (3.9.1) as identified.

CHAPTER 4

CALCULATION OF THE VARIABLES

4.1 Introduction

As important in determining the results of this analysis as the specification of the functional form are the details of the definitions of the variables used. We therefore devote this chapter to describing, for each variable in turn, the derivation of its precise definition from theoretical postulates and empirical evidence and the calculation of the time-series, and to indicating the likely sources of errors in the series and their effect on the estimated coefficients.

Although deficiencies of the data have necessitated the introduction of many statistical assumptions, it has been possible to obtain series which are better approximations to the theoretical requirements than the most readily available summary statistics, and to obtain Area figures for variables for which none are given in the main official publications. For each variable included in the regressions a set of twelve time-series covering 52 quarters has been calculated, giving 624 observations in all.

The calculations described are those for the twelve Electricity Board Areas in England and Wales in the period 1955II to 1968I. The calculations on South of Scotland data, to be used for predictive testing, and those to extend the time-series back to 1951 or 1952, to permit inclusion of lagged variables, were similar.

4.2 The definition of the dependent variable

Theoretically the dependent variable should be the total quantity of electricity demanded per household.

The sum of the dependent variables in our unrestricted and off-peak analyses,[1]

$$Q \equiv Q^u + Q^o,$$

is the quantity of electricity purchased from the Electricity Boards on their domestic unrestricted tariffs (Q^u) and domestic off-peak tariffs (Q^o). C is the

(1) The latter described in Chapter 7.

number of persons billed. Hence Q/C differs from the theoretical definition because:

(1) Consumption of supplies from batteries and private generators is excluded. Whilst this discrepancy may be important for some types of appliances within the sundry class, for example radios, the proportional error in the total for the class is negligible.

(2) Some households in institutions, hotels, combined domestic and commercial premises, and farms were supplied on other tariffs. This should make a negligible difference to the *mean* consumption per consumer.

(3) The number of households did not equal the number of persons billed, or "electricity consumers", firstly because not all dwellings were wired for a mains supply, and secondly because some bills were for supplies to more than one household. Needleman[2] estimated that the proportion of households wired in the United Kingdom rose from 88% in 1955 to 96% in 1968. As the figures for England and Wales were probably higher, we would not expect large errors in the *mean* values of any of the exogenous variables. The incidence of multi-household consumers only affects the total consumption expenditure variable, E—the appliance data are for consumers. The Electricity Council estimated that they were 1% of all consumers in England and Wales in 1966, and that the proportion was highest in the London Area at 5%. No correction has been made.

4.3 The difference between units billed and units consumed

The series for Q^u gives not units consumed in each calendar quarter but units billed, that is, recorded as having been consumed since the previous meter-reading on meters read in the quarter. Since meters were normally read at regular quarterly intervals, and meter-reading was spread over the whole quarter, units billed to all consumers in quarter q include only about half of their consumption in that quarter and about half of their consumption in quarter $q-1$.

Let Q_q^u be units billed in quarter q,
   D^m the number of days in month m,
   M_q^m the proportion of month m's consumption read in quarter q,

and quarter q consist of months $3q-2$, $3q-1$, $3q$.

Then
$$Q_q^u = \sum_{m=3q-5}^{3q} M_q^m \sum_{d=1}^{D^m} Q_{dm}^u \tag{4.3.1}$$

We shall refer to this as consumption in *moving-quarter q*.

(2) Needleman (1960) p. 40.

4.4 Aggregation over time

Aggregating the demand function (3.9.1) over time by applying the same ranges of summation as in (4.3.1) yields

$$\sum_m M_q^m \sum_d \left(\frac{Q_{dm}^u}{C_{dm}}\right) = \beta_1 \sum_m M_q^m D^m + \sum_m M_q^m \sum_d \sum_a Z_{dm}^a \sum_{j=2}^{11} \beta_{j,a} X_{dmj} + \sum_m M_q^m \sum_d \varepsilon_{dm}$$

(4.4.1)

Given that changes within months in C_{dm} and Z_{dm}^a were very small, no significant error is introduced if the following assumptions are made:

Assumption 4.4.1

$$\sum_m M_q^m \sum_d \left(\frac{Q_{dm}^u}{C_{dm}}\right) = \frac{\sum_m M_q^m \sum_d Q_{dm}^u}{C_q}$$

(4.4.2)

where

$$C_q \equiv \frac{\sum_m M_q^m D^m \left(\sum_d C_{dm}/D^m\right)}{\sum_m M_q^m D^m}$$

(4.4.3)

the mean number of consumers in moving-quarter q.

Assumption 4.4.2

$$\frac{\sum_d Z_{dm}^a X_{dmj}}{D^m} = \frac{\sum_d Z_{dm}^a}{D^m} \cdot \frac{\sum_d X_{dmj}}{D^m} \quad \text{all } j, a$$

(4.4.4)

Then, writing

$$\lambda_q^m \equiv \frac{M_q^m D^m}{\sum_m M_q^m D^m} \qquad m = 3q-5, \ldots, 3q$$

(4.4.5)

$$Z_q^a \equiv \sum_m \lambda_q^m \frac{\sum_d Z_{dm}^a}{D^m}$$

(4.4.6)

$$X_{qj} \equiv \sum_m \lambda_q^m \frac{\sum_d X_{dmj}}{D^m}$$

(4.4.7)

$$\varepsilon_q \equiv \sum_m \lambda_q^m \frac{\sum_d \varepsilon_{dm}}{D^m}$$

(4.4.8)

and the mean quantity demanded per consumer per day in moving-quarter q, in kilowatt-hours, as

$$Y_q^u \equiv \frac{Q_q^u}{C_q \sum_m M_q^m D^m}$$
(4.4.9)

the aggregate demand function (4.4.1) becomes

$$Y_q^u = \beta_1 + \sum_{a=1}^{4} Z_q^a \sum_{j=2}^{11} \beta_{j,a} X_{qj} + \varepsilon_q$$
(4.4.10)

This is the final form to be estimated, subject to constraints on the parameters introduced in Chapter 6.

4.5 The error variance-covariance matrix

Given assumption 3.10.1 on the homoscedasticity and independence of ε_{dm}, the error ε_q is heteroscedastic and successive terms ε_q and ε_{q+1} have a positive covariance.

Writing

$$E\{\varepsilon_B \cdot \varepsilon_B'\} = \sigma^2 \mathbf{V}_B$$

where $\varepsilon_B = \|\varepsilon_q\|$ is the (52×1) vector of errors for Area B, $\mathbf{V}_B$ is a (52×52) matrix with elements

$$
\left.
\begin{aligned}
v_{q,q+k} &= \sum_{m=3p-5}^{3q} \frac{(\lambda_q^m)^2}{D^m} & k&=0 \\
&= \sum_{m=3q-2}^{3q} \frac{\lambda_q^m \lambda_{q+k}^m}{D^m} & k&=1 \\
&= 0 & k&>1 \\
&= v_{q+k,q}
\end{aligned}
\right\}
$$
(4.5.1)

4.6 Calculation of the monthly weights, λ_q^m

A first approximation to the monthly weights λ_q^m may be obtained by making the following assumptions:

Assumption 4.6.1

One-third of consumers' meters are read each month.

Assumption 4.6.2

Meter-reading takes place at a constant rate per day within each month.

Assumption 4.6.3

Electricity is consumed at a constant rate per day within each month.

These give the same values M_q^m and λ_q^m for all q:

m	M_q^m	λ_q^m
$3q-5$	0·1667	0·0556
$3q-4$	0·5000	0·1667
$3q-3$	0·8333	0·2778
$3q-2$	0·8333	0·2778
$3q-1$	0·5000	0·1667
$3q$	0·1667	0·0556

Assumptions 4.6.1 and 4.6.2 are implausible in view of the variations in the number of working days in a month and the postponement of some meter-readings due to the consumer's being out when the meter-reader first called. Alternative assumptions which take account of these factors were made when calculating the weights finally used to compute the moving-quarter values C_q, X_{qj}, and Z_q^a and the elements of $\mathbf{V}_B$. These assumptions give a different set of weights for each quarter. The average pattern for each quarter of the year is shown in Table 4.6.1. The most important result of the change in the assumptions is to give relatively more weight to months in the previous quarter.

TABLE 4.6.1. *The monthly weights,* λ_q^m
Means for the period 1952II-1968I

Month	Quarter			
	II	III	IV	I
$3q-5$	0·0608	0·0546	0·0611	0·0600
$3q-4$	0·1618	0·1777	0·1793	0·1728
$3q-3$	0·2929	0·2870	0·2860	0·2943
$3q-2$	0·2761	0·2801	0·2770	0·2773
$3q-1$	0·1643	0·1600	0·1535	0·1476
$3q$	0·0441	0·0406	0·0430	0·0481

Assumption 4.6.3 has not been relaxed. It is invalid because:

(1) There are large irregular changes in consumption within months due to changes in natural conditions; and

(2) Temperature and hours of daylight are subject to trends within each month.

Adjustment for the irregularities is not feasible in this study since lengthy analysis of daily consumption and meteorological data would be required. As a result the explanatory power of the temperature variable will be lower.

On estimating the size of the correction to take account of the trend, it was found not to be large enough to justify the increase in computations required.

4.7 Calculation of the dependent variable, Y^u

The number of consumers was available only as an annual series. But because of the low rate of change, the mean moving-quarter value, C_q, may be satisfactorily estimated from linearly interpolated monthly figures.

For the quarters before off-peak rates became significant, off-peak units billed (Q^o) had to be estimated, but the possible resulting error in unrestricted units billed (Q^u) is negligible. For later quarters Q^u was available directly from records on the whole population of consumers.

Thus the series for the dependent variable Y_q^u defined by (4.4.9) are very accurate.

4.8 Ownership of space-heating and water-heating appliances— Z^1, Z^2, Z^{o1}, Z^{o2}

Appliance class 1 is defined as unrestricted space-heaters comprising all direct-acting space-heaters; and class 2 as unrestricted water-heaters, comprising those immersion heaters, large self-contained or storage heaters, and sink storage and non-storage heaters on unrestricted tariffs.

The national average installed load of space-heaters per consumer having, l^1, rose between 1955 and 1966 from 2·35 kilowatts to 3·8 kilowatts, and the installed load of water-heaters, l^2, from 2·4 kilowatts to 2·9 kilowatts. The figures for individual Areas varied greatly for space-heaters, from 3·1 to 4·5 kilowatts in 1966, but very little for water-heaters.

Our hypothesis is that increases in l^1 and l^2 were both caused by increases in the demand for heat from these appliances rather than a desire to meet the same demand in a shorter time. Accordingly the ownership variables are defined as follows:

$$Z^1 \equiv (Al)^1$$

the installed unrestricted space-heating load per consumer;

$$Z^2 \equiv (Al)^2$$

the installed unrestricted water-heating load per consumer.

Such an increase in demand could arise for any of the following reasons: an increase in the demand for heat; a shift in the demand for heat from other appliances, such as kettles, to appliances in these classes; an increase in the proportion of owners of these classes of electrical appliances who use them as their main form of heating—for example associated with the decline in the use of solid fuel—or who make relatively frequent use of electrical appliances for supplementary heating.

To the extent that heaters were acquired to be used only infrequently, or to achieve a given temperature more quickly, the definitions used give rise to negative Z-U effects. But if an increased demand for heat led to increased utilization as well as a higher installed load, there would be positive Z-U effects.

The off-peak ownership variables are defined as follows:

$$Z^{o1} \equiv (Al)^{o1}$$

the installed off-peak space-heating load per consumer;

$$Z^{o2} \equiv (Al)^{o2}$$

the installed off-peak water-heating load per consumer.[3]

4.9 Ownership of cooking appliances—Z^3

Appliance class 3 is defined as all full-size and breakfast-type cookers.

There were only small variations in l^3 between Areas. From 1955 to 1961 the national average fell slightly, from 6·0 to 5·9 kilowatts, as full-size cookers increased in size but the proportion of the smaller breakfast-type rose. From 1961 to 1966, l^3 rose to at least 7·7 kilowatts, on the assumption that the load per cooker rose by at least as much as the average number of hotplates. This was only in small part due to a decline in the proportion of breakfast-type.

As it was rare to own two electric cookers, l^3 measures the rating of a single cooker for an individual household. It was also rare to own both an electric and an other-fuel cooker. Some owners of electric cookers used a gas ring—and the proportion doing so declined—but the proportion using an electric kettle rose. Hence the demand to be met by an electric cooker can on average have risen little unless the demand for cooking facilities rose. It seems unlikely that the average demand from all households rose much over the period 1955 to 1968, with the fall in household size and increasing consumption of convenience foods. It appears that at least in part the rise in ratings was due to the provision of more special facilities on new cookers. One hypothesis is therefore that the variations in l^3 were due to a desire to meet the same demand in a shorter time and hence that the correct variable to use is

$$Z^3 \equiv A^3$$

the proportion of consumers owning electric cookers.

It is, however, possible that households with a relatively high demand for cooking facilities were switching to electricity and that this explained the rise in the number of hotplates per cooker, but not any increase in the rating per hotplate. As l^3 had to be calculated on the assumption that it changed in proportion to the number of hotplates per cooker, this implies using:

$Z^3 \equiv (Al)^3$, the installed load of cookers per consumer.

(3) The calculation of these is described in Chapter 7.

Another hypothesis is that variations in the proportion of breakfast-type were caused by variations in the average demand for cooking facilities from owners of electric cookers but that variations in l^3 for other reasons are explained by our first hypothesis. This implies using

$$Z^3 \equiv (A\tilde{l})^3$$

where $(A\tilde{l})^3$ is a weighted average of the proportion of consumers owning each type of cooker using the same weights for all periods. Weights of 6·5 kilowatts (full-size) and 3·0 kilowatts (breakfast-type) have been used as these were the figures for 1961 and were also very close to the figures for earlier years.

All three variables were calculated; the results with each are compared in Chapter 6. Except where stated otherwise

$$Z^3 = A^3$$

If an increase in efficiency accompanied the increase in ratings and ownership, then the omission of any variable measuring efficiency will lead to the coefficients of other variables taking up a negative Z-U effect.

4.10 Ownership of lighting and sundry appliances—Z^4, Z^{41}, Z^{42}

Appliance class 4 is subdivided into two classes:

class 41—lighting, comprising filament and fluorescent lamps;
class 42—sundry appliances, comprising all appliances not elsewhere included.

Because the sundry appliances meet very different demands, and consumers would not normally have had more than one of most types, changes in the individual consumer's installed load of them would not have been associated with changes in utilization as under assumption 3.3.2(ii) on the independence of A and lur unless additions to the stock were mainly appliances with the lowest rates of utilization per kilowatt. Accordingly the ownership variable is defined as

$$Z^{42} \equiv (Al)^{42}$$

Because of the limited coverage of surveys in the earlier years, this has been calculated as the installed load of kettles, washboilers, washing machines, irons, refrigerators, mains radios, television sets, and dishwashers, assuming that the average rating of each type remained constant. Variation in this series should be representative of variations in the total installed load of sundry appliances since the appliances listed accounted for over 75% of the total. No highly utilized appliances are omitted, so that those listed accounted for much more than 75% of the demand due to use of sundry appliances. There should therefore be only a small and constant proportional error in $(Al)^{42}$ and over-estimate of $(ur)^{42}$

As it is necessary to use

$$Z^4 \equiv Z^{41} + Z^{42}$$

Z^{41} is defined as

$$Z^{41} \equiv (Al)^{41}$$

the installed lighting load per consumer, calculated on the assumption that the average rating per lamp, of both kinds, remained constant at 70 watts, as the data on the size of bulbs are unreliable.

The national average number of filament lamps per consumer rose from 9·0 in 1955 to 9·8 in 1961 and 11·3 in 1966, and the average number of fluorescent lamps also rose. The implied hypothesis is that this rise was due to provision of supplementary lighting in some rooms which was used as much as the main lighting. To the extent that it was due to the provision of less-used lamps, there will be a negative Z-U effect on U^4, but this could be cancelled out if the average wattage of all lamps rose.

4.11 The reliability of the estimates of appliance ownership

The main sources for all the ownership series are the Electricity Council surveys of 1955, 1961, and 1966. These were sample surveys: for a proportion of consumers owning, A, the 95% confidence interval assuming only a random sampling error, ranges from $\pm 0·044$ if $A = 0·500$ in the smallest Area to $\pm 0·020$ if $A = 0·900$ in the largest; and there may have been selection bias, in particular that arising from non-response, necessitating the use of wider intervals. Installed load figures are subject to even wider margins of error given the difficulties of collecting the required information on wattage ratings from households and the need to use guesses for several series.

For lighting and sundry appliances, the three observations obtained from these sources for each Area were the only ones available. For the other classes there were other surveys giving up to eleven more observations for each Area. The estimates obtained from these surveys are not, however, consistent with those from the Electricity Council surveys. As they should give better estimates of year-to-year changes, on average, than an assumed trend, they have been used after adjustments to achieve consistency. All the other quarterly values were obtained by interpolation or extrapolation.

Despite the deficiencies of the series so obtained, much improved results can be expected compared with those achieved using only a national time-series with assumed trends. For although the trend in individual Areas over short periods is open to considerable doubt, trends over the whole period and most inter-Area differences are estimated with reasonable accuracy.

4.12 The availability of a gas supply—G

In section 3.8 consumers were divided into those who would regard themselves as having a gas supply available at negligible cost (a proportion G of the total) and those who would not.

The Electricity Council survey of 1966 gives the proportion having no gas pipes anywhere in the house. Of these consumers, some would probably not have been deterred from acquiring a gas appliance by the cost of getting a supply. On the other hand, some with pipes may have been deterred from acquiring a particular appliance by the cost of getting pipes at the right place. On the assumption that the difference between the numbers in these two categories is small, G is defined as the proportion having gas pipes somewhere in the house.

As there are no figures for other years it is assumed that G is invariant over the period under study in each Area. As it would change little unless the proportion of new housing having a supply was very different from that of the existing stock, this is probably a very good approximation. Nationally the change in G appears to have been small: in 1966 G was 0·76 in England and Wales and in 1955 was at most 0·09 below this since the ownership level of gas cookers was 0·67.

4.13 Evidence on the relationship between ownership of electrical appliances and availability of a gas supply

The zero-order correlation coefficients between G and the other exogenous variables are all very low absolutely, except that $r(G, \text{lag}_0 \ln(\bar{E}/\bar{\pi})) = 0\cdot54$. For space-heaters, water-heaters, and lighting and sundry appliances, the correlation between G and the ownership level Z^a is also very low. Together this information suggests that, although the equality of assumption 3.8.2(ii), $Z^{aG} = Z^{aNG}$, did not hold, the ratios Z^{aG}/Z^a and Z^{aNG}/Z^a which appear in (3.8.1) were nearly constant and probably close to unity. For lighting alone, the equality must have held almost exactly since there were practically no gas substitutes.

For cookers the evidence of a relationship between ownership and gas availability is much stronger: $r(Z^3, G) = -0\cdot63$. There are sufficient data to enable us to put bounds on Z^{3G} and Z^{3NG}. On the assumption that no household owned both an electric and a gas cooker, the upper bound to Z^{3G} is obtained by assuming further that all other cookers were owned by those without a gas supply, and the lower bound by assuming that they were owned by those with a gas supply. These bounds have been calculated for England and Wales in 1955, 1961, and 1966, and for each Area in 1955, when they were widest apart.

The results show Z^{3G} to have been positive and well below Z^{3NG} in all

Areas and all years, invalidating both alternatives within assumption 3.8.2. Most of the Area values of the ratios Z^{3G}/Z^3 and Z^{3NG}/Z^3 were, however, clustered together: only one ratio Z^{3G}/Z^3 (for London) was definitely outside the range $0\cdot50\pm0\cdot05$, and only two ratios Z^{3NG}/Z^3 (for London and South Western) outside $2\cdot54\pm0\cdot30$. The national figures do not refute the hypothesis that there was little trend in the ratios.

Thus it probably does not introduce too serious an error into (3.9.1) to assume that $\beta_{10,a}$ and $\beta_{11,a}$ are constants for all a.

4.14 The price of electricity—P^e

In the period under study, standard domestic unrestricted tariffs were of two types: either "two-part", that is, a fixed charge, usually related to the size of dwelling, and a flat rate; or "block", that is, a set of two or three marginal rates giving marginal price as a decreasing step function of the quantity supplied, the position of the steps often depending on the size of dwelling. At certain times some Boards had more than one such standard tariff and all Boards offered other special rates.

Typically, however, a Board had one standard tariff which gave the minimum average price for all consumption levels, and almost all consumers were on this tariff. There may have been some who were not on the most economical tariff for their consumption level because of imperfect knowledge or because they were tenants of the person billed and not free to choose, but the following assumption should be realistic for most consumers:

Assumption 4.14.1

Let $p_q^h = p_q^h(Q_q^{uh})$ give the marginal rate on the tariff for household h which minimizes the average price at consumption level Q_q^{uh}, and let this be called the *tariff function*. Then the tariff function gives the marginal rate actually paid.[4]

The subscript q and superscript h are suppressed for the rest of this section.

To simplify analysis of the effects of changes in the tariff function, it is assumed that there was no uncertainty about expected future prices; that any changes in the tariff function were expected to be permanent; and that the prices of other commodities were given independently of the quantity

(4) Prepayment customers paid a surcharge which was sometimes a fixed charge per quarter and sometimes took the form of an addition to the primary rate (but not the final rate). The minimization is interpreted as subject to the type of meter installed, so that in the latter case prepayment and credit customers had different tariff functions. Some tenants paid more than the Electricity Board's rate. Surcharges imposed by landlords have had to be ignored in subsequent calculations as no information was available.

demanded. Relaxation of these assumptions does not necessitate any modification to our conclusion about the price variable in the utilization function.

Since, for most households, expenditure on electricity was a small proportion of total consumption expenditure, we may reasonably make

Assumption 4.14.2

The income effect of a change in the tariff function on the demand for
electricity was negligible.

Let $P^e = P^e(q^u)$ be the (monotonic decreasing) function relating the current
demand price of electricity, P^e, to the quantity demanded by household h
in a long-run (appliance stock and utilization) equilibrium, holding income,
tastes, and the prices of all other commodities constant, and supposing that
the same rate is paid for all units consumed.

Given assumption 4.14.2, if the long-run equilibrium quantity demanded
is q, and $p*$ is the marginal rate at this consumption level,

$$p* = P^e(q*) \qquad (4.14.1)$$

irrespective of whether this rate $p*$ is paid for *all* units consumed or not.

Because of the steps in the tariff function, there may be more than one
price satisfying

$$P^e = p \qquad (4.14.2)$$

The equilibrium price $p*$ is that corresponding to the quantity $q*(\geqslant 0)$
which maximizes

$$\int_0^{q^u} (P^e - p)\, dq^u \qquad (4.14.3)$$

Let an initial equilibrium be within the consumption range for which the
tariff function gives the marginal rate as p^1. Then the following results may
be derived:[5]

(1) if p^1 changes, then $q*$ *and* $p*$ change;
(2) if p^1 remains the same and other rates change then either $p*$ and $q*$
    remain the same if the new value of (4.14.3) at the value of q^u in the
    initial equilibrium is still the maximum, or both $q*$ and $p*$ change
    if (4.14.3) now attains a maximum for some other q^u.

These results apply to all tariffs. A two-part one may be treated as
formally equivalent to a two-block tariff with a single-unit primary block.

Thus, in an expression for the long-run equilibrium demand of an individual household, the current price variable should be the marginal rate

---

(5) The case where initially $q* = 0$ provides exceptions. For simplicity these are
omitted as they do not alter the conclusion reached.

given by its tariff function at the equilibrium level of consumption; and, in the aggregate function, an average of these marginal rates weighted in each utilization function U^a by the corresponding ownership levels Z^{ah}.

The actual demand of a household in any quarter may depend on a different marginal rate if expectations about the other determinants of demand held at the beginning of the quarter are not fulfilled or if demand does not adjust to regular seasonal variations in the marginal rate actually paid. Which is the relevant rate in these circumstances will be discussed in Chapter 8 where an average of the marginal rates is to be used in the aggregate demand function.

In this model, for reasons already given in 3.11, the final rate on the standard domestic unrestricted tariffs, denoted P^{ef}, is used. As it was directly available, the only error in the variable arises from the difference between P^{ef} and the appropriate average of the rates.

These should be highly correlated, firstly because in most years most consumers would have expected to have, and did have, P^{ef} as their marginal rate in all quarters. In 1955, 15% of consumers had the primary rate as their marginal rate in the summer, and 4% in the winter. The figures would almost certainly have declined over the subsequent period as the size of primary blocks tended to fall as a proportion of mean consumption. Secondary blocks in three-block tariffs were often large, but an Area Board had a three-step tariff function on average in only 12 quarters between 1955ɪɪ and 1968ɪ. Secondly the correlation should be high because those having a pre-final rate as their marginal rate would tend to be those with relatively low appliance stocks, Z^{ah}.

4.15 The price of gas—P^g

The tariff function for gas was also a decreasing step function of consumption. For many consumers a change in gas consumption would have necessitated a change of tariff in order to obtain the minimum rate for their new consumption level given by the tariff function. Hence because of lagged response and imperfect knowledge, the proportion of consumers having a marginal rate above that given by the tariff function was probably higher for gas than for electricity, particularly from 1961 onwards when mean gas consumption was rising fairly fast. Thus assumption 4.14.1 needs to be relaxed for gas, but otherwise arguments similar to those in 4.14 apply: the gas price variable in the utilization function U^a should be an average of the marginal rates actually payable by each electricity consumer weighted by the corresponding ownership levels Z^{ah}.

Lack of data on the distribution of gas consumption about the mean precludes calculation of the weighted average. It is unsatisfactory for gas to use the final rate since the correlation between this and the weighted average would be much lower than for electricity: the shape of the tariff function varied more and there must have been associated changes in the proportion of consumers on each rate; a relatively low proportion of gas consumers were on the final rate, and would have tended to be those with low stocks of electrical appliances, at least in the space-heating, water-heating and cooking classes.

As a better approximation, $\bar{P}^g$ has been calculated as the average marginal rate given by the tariff function for credit customers using town gas over the range from 50% of current mean consumption to 250%. This includes the consumption levels used to calculate "typical retail prices" by the Ministry of Power.

The only figures available for mean gas consumption were for years ending 31 March. The range for any given quarter has therefore been calculated from mean consumption in the year centred at the 30 September occurring in the year ending at the centre date of the quarter, on the assumptions that utilization of electrical appliances did not respond to regular seasonal variations in prices,[6] and that, if a change of tariff was required to obtain the minimum rate, then this would be made so that all consumption from the middle of quarter IV onwards was on the new tariff. The rationale is that most changes would have been to tariffs which were more economical only at higher consumption levels, first reached during the winter. The assumptions make little difference to lag ln $(\bar{P}^g/\bar{\pi})$.

It was not possible to adjust the annual figures to normal weather conditions. The result is that the calculated series for $\bar{P}^g$ falls too much, *ceteris paribus*, when there was an abnormally severe winter, such as that of 1962–63, since an abnormally high proportion of consumers would then not have been on the cheapest rate for their unexpectedly high consumption level. The resulting bias towards zero in the coefficient of lag ln $(\bar{P}^g/\bar{\pi})$ should, however, be negligible since the correlation between it and the temperature variable in the 624 observations was only 0·0741.

Gas Board and Electricity Board Areas did not coincide and Gas Board Areas were subdivided into tariff zones. The proportion of electricity consumers in each Gas Board zone–Electricity Board Area intersection was estimated from data on the distribution of electricity consumers between the Electricity Board Districts within each Area. A rough adjustment was then made for the difference between this distribution and that of those with a gas supply, by increasing the weight given to relatively urban intersections. Some error in the estimation of inter-Area differences in the mean price may have resulted but should have diminished over the period as standardization of gas tariffs within Gas Board Areas proceeded.

(6) See also 4.19 below.

4.16 The price of solid fuel—P^c

The main problem in constructing series for "the price of solid fuel" is to get a representative measure of changes in the prices of the great variety of solid fuels. The merchants' price for a particular kind did vary with the size of order but only by a small amount. It is here a good enough approximation to use the same marginal price for all consumption levels.

The series were constructed by using the price of house coal (Group IV) delivered by merchants in the Decembers of 1964-68 in one town in each Area, and interpolating and extrapolating all other monthly values on the assumption that the trend up to December 1964 and the monthly variations in each Area were the same as those indicated by the Index of Retail Prices Coal and Coke sub-index for the United Kingdom.

Some error therefore arises because (1) the one town for which there are price data was not representative of the whole Area, and (2) there were variations in house-coal consumption as a proportion of solid fuel consumption and in the price of house coal relative to other solid fuels. But the largest error arises because the merchants' price is not the price paid by all consumers. For example at the end of our period, at least a quarter of domestic coal consumption in the South Wales Area was miners' free and concessionary coal. As regional variations in merchants' prices were largely determined by transport costs, the result of the errors is that the calculated series probably give a correct ranking of Areas but understate inter-Area differences in the unweighted mean price of solid fuel, as given by (3.4.4). The underestimation of the difference in the weighted means is probably less, owing to positive correlation between stocks of electrical appliances, Z^{ah}, and the price of solid fuel within Areas, but the estimated absolute values of the coefficients of the price of solid fuel in the utilization functions U^a will probably be too high.

4.17 Total consumption expenditure per household—E

There are three parts to our hypothesis about the income determinant of utilization.

Firstly, in a world with a perfect capital market and perfect knowledge of future income, a consumer's total current consumption expenditure (E), and utilization and ownership of appliances would depend on some quantity such as "permanent" or "normal" income,[7] say I^*; current income, I, would affect them only through its effect on I^*. The effect of imperfect knowledge is that changes in I lead to revised expectations of future income, changes in I^* and hence changes in utilization. The effect of imperfections in the capital market is that I is a constraint on the attainment of the desired utilization (and more especially of the desired stock of appliances). But both imperfections also affect E in the same way. Therefore E is the best single

(7) See Friedman (1957) and Farrell (1959).

variable to use in the utilization functions.

Secondly, the decision-making unit is the household, so that total household consumption expenditure and utilization do not depend on the number of income-recipients. In the first part of the hypothesis, "the consumer" is the household.

Thirdly, an increase in household size has a *scale effect* on utilization which cancels out a significant part of the *income effect*, where by scale effect we mean the effect of a change in household size holding income per person constant, and by income effect the effect of a change in income per person, holding household size constant.[8] Therefore, given that household size is not included as a separate variable, mean total consumption expenditure is better expressed per household than per head.

The variable E is defined as the mean expenditure per household on Item Groups 1-94 in the Family Expenditure Survey.[9]

The data refer to Standard Regions instead of Electricity Board Areas; were available only for 1953 and 1961-67, annually; and were obtained from a sample—and one not designed before 1967 to give regional figures—so that there are errors in the regional figures due to both random sampling fluctuations and regional imbalance.

To estimate the proportion of the consumers in each Electricity Board Area who were in each Standard Region–Area intersection, 1961 population figures were used. It was assumed that household size did not vary within Areas between intersections; and that the rates of growth of population were the same in each intersection within an Area, except that allowance was made for the decline in Greater London's population being due entirely to a decline within the London Electricity Board Area.

Both means and standard errors were available for E in each Standard Region. To obtain variances for Areas it was assumed that the error for each Standard Region intersecting a given Area was independent. This would not be true for all the Standard Regions in the United Kingdom together because errors due to regional imbalance would cancel out, but is satisfactory for small groups of Regions. To obtain first estimates of Area means it was assumed that the mean for each intersection within a Standard Region was the same. First estimates $\hat{E}_{B,t}$ were obtained for the years 1953, 1961-67 ($t = -7, 2, ..., 7$).

To attempt to separate variations in $\hat{E}_{B,t}$ due to sampling errors from those due to variations in the true figure $E_{B,t}$, it was assumed that an Area figure was related to the Survey United Kingdom mean $\hat{E}_{UK,t}$ by

$$\frac{\hat{E}_{B,t}}{\hat{E}_{UK,t}} = \alpha_1^B + \alpha_2^B t + \eta_{Bt} \qquad (4.17.1)$$

(8) See Prais & Houthakker (1955) chapter 10.
(9) A definition virtually identical to that used in Stone (1954), as quoted in 2.2.

where η_{Bt} is a random error with zero expectation and subject to

$$\sum_{B=1}^{12} \eta_{Bt} N_{Bt} = 0 \quad \text{all } t \tag{4.17.2}$$

where N_{Bt} is the sample size—the existence of a sampling error in the Survey England and Wales mean, $\hat{E}_{EW,t}$, is ignored as it is much smaller than the errors in the Area figures. On the assumption that the variance of η_{Bt} is proportional to $(\hat{E}_{EW,t})^2/N_{Bt}$, the parameters α_1^B and α_2^B may be estimated by applying Weighted Least Squares to the time-series for each Area ignoring (4.17.2).

As a check on the assumptions, the England and Wales mean for each year was computed from the regressions and compared with the Survey figure. The agreement was close for all years.

Estimates of the United Kingdom mean, $E_{UK,t}$ for years other than Survey years were obtained using data on consumers' expenditure in *National Income and Expenditure, 1969* and the London and Cambridge Economic Service's *The British Economy: Key Statistics 1900-66* and on the number of households in Needleman (1960). Final estimates of the annual Area means were then obtained from the regressions. The derivation of the required quarterly series for $\bar{E}/\bar{\pi}$ is described in 4.19 below.

Because irregular and cyclical movements in the true ratios $E_{B,t}/E_{UK,t}$ and sampling errors in $\hat{E}_{UK,t}$ have been ignored, there are probably errors in the series but they should not be so large as to introduce a serious bias into our estimates of the coefficients in the demand function.

Because E varied within Areas and lagged values were probably important determinants of appliance ownership, the calculated unweighted mean differs from a mean weighted by the ownership levels Z^{ah}, violating assumption 3.4.2. This means firstly that there are (possibly large) Z-U effects to be explained by the coefficients of lag $\ln(\bar{E}/\bar{\pi})$, $\beta_{7,a}$: if the distributions of tastes and income were such that an increase in the income of all households caused a relative increase in appliance ownership amongst poorer households, the weighted mean could actually fall; if there was no associated change in the weighted mean of any other variable and the sum of the U effect and any other Z-U effect is positive, this implies a negative Z-U effect. On the other hand, if the use of gas when it is available is negatively correlated with E, then the positive inter-Area correlation between the availability of a gas supply, G, and lag $\ln(\bar{E}/\bar{\pi})$ implies a greater difference between Area weighted means than between the unweighted means, and so a positive Z-U effect.

Secondly it means that the coefficients of other variables include Z-U effects which would be included in $\beta_{7,a}$ if a weighted mean were used for E. For example, the coefficients $\beta_{9,a}$ of the price of solid fuel include a positive Z-U effect if (1) $\beta_{7,a} > 0$ and (2) where the price of solid fuel, $\bar{P}^c$, is low, ownership of electrical appliances is lower particularly among high-income households, or the distribution of income is more positively skewed—so that $\bar{P}^c$ is positively correlated with the difference between the weighted and unweighted means of expenditure.

4.18 The index of all other prices—π

If utilization U^a were a linear function of the logarithms of E, P^e, $\bar{P}^g$, $\bar{P}^c$ and the prices of all other commodities, say $p_1, p_2, \ldots, p_{N-3}$; if the function were homogeneous of degree zero, and the cross-elasticity with respect to $p_i(i=1, 2, \ldots, N-3)$ proportional to expenditure on commodity i in some year t; then the deflator of E, P^e, $\bar{P}^g$ and $\bar{P}^c$ should be a geometric mean of $p_1, p_2, \ldots, p_{N-3}$ weighted by expenditures in year t.

Given that the cross-elasticities and form of U^a may not satisfy these conditions and that expenditure on fuels formed only a small part of total consumption expenditure in our period, it is satisfactory to use the official Index of Retail Prices, even though this is an arithmetic mean of all prices including fuel prices and the series for 1955-68 is of the chain-linked form.

In the absence of any reliable information on variations in the price level between regions, it is assumed that there was none. Thus for each Area, π is the Index of Retail Prices for the United Kingdom, with 16 January 1962 = 1.

If the Areas with the highest consumption expenditure per household—London, Southern, South Eastern, and Eastern—also had relatively high price levels, the estimated coefficient of $\bar{E}/\bar{\pi}$ will tend to be too close to zero.

4.19 Lagged variables

Given the data available and the form of the model, it is not feasible to include a dozen or more lagged values of total consumption expenditure and the fuel prices in order to estimate their coefficients separately. It is, however, possible to calculate lag ln $(P^{ef}/\bar{\pi})$, lag ln $(\bar{P}^c/\bar{\pi})$ and lag ln $(\bar{P}^g/\bar{\pi})$ using several dissimilar sets of weights and then compare the regression results.

The data on the prices allow us to have up to a three-year lag without losing any observations in the period from 1955II onwards. As it is unlikely that the U effect of a price change would be spread over more than three years, the weights are restricted to add to unity over twelve quarters including the current one: let w_t be the weight in moving-quarter t, numbering from the current moving-quarter as 0. Then

$$\text{lag } X_{qj}^* = \sum_{t=-11}^{0} w_t X_{q+t,j}^* \qquad j=8, 9, 10 \qquad (4.19.1)$$

where
$$\sum_{t=-11}^{0} w_t = 1 \qquad (4.19.2)$$

$$X_{q8}^{*} = \ln (P^{ef}/\bar{\pi})_q \qquad (4.19.3)$$

$$X_{q9}^{*} = \ln (\bar{P}^{c}/\bar{\pi})_q \qquad (4.19.4)$$

$$X_{q10}^{*} = \ln (\bar{P}^{g}/\bar{\pi})_q \qquad (4.19.5)$$

A simple hypothesis is that the aggregate adjustment path is linear over three years. The variables (4.19.1) are calculated on this hypothesis as a basis for comparison with the other hypotheses and written

$$\left. \begin{aligned} \text{lag}_0\, X_{qj}^{*} &= \sum_t w_t^0 X_{q+t,j}^{*} \qquad j = 8, 9, 10 \\ w_t^0 &= \tfrac{1}{12} \qquad t = -11, \ldots, 0 \end{aligned} \right\} \qquad (4.19.6)$$

where

As expenditure on fuel was typically a small proportion of the total consumption expenditure of a household, a rise in a fuel price would not necessitate an immediate change in fuel consumption as long as small changes in the total budget could be accommodated. Given the strength of habits in utilization, and the need possibly for a change in the behaviour of members of the household other than those who would normally be involved in budget decision-making, the initial reaction to a perceived change in price might well be slow. Credit electricity and gas consumers might not perceive a change immediately and, given the complexity of the tariffs, might only slowly perceive its exact effect on expenditure. Thus there is unlikely to be a continuously declining rate of adjustment. More plausibly the adjustment path would be skew and sigmoid, so that on aggregation a skew sigmoid curve would be obtained. Possibly some consumers had a non-monotonic path if they overreacted on perceiving a change. As long as they were few in number or took a varying time to react, the aggregate path would still be of the same general shape.

As there is no information on which to base a more precise specification of the adjustment path, for convenience the two-parameter cumulative lognormal curve is used.[10]

To obtain a set of weights two conditions on the parameters are required: as the first, the integral of the lognormal density function over the range 0 to 12 is restricted to be 0·995. The integral over each unit interval is then multiplied by 1/0·995 to give a set of weights fulfilling (4.19.2). This procedure is adopted as it is not possible to restrict the integral to unity over a finite range; the choice of 0·995 is arbitrary but the weights are insensitive to small variations.

As the second condition the mode is specified, since it has a simpler

(10) See Aitchison & Brown (1957).

interpretation than the moments. The variables were calculated for two different sets of lognormal weights representing what appear to be the most plausible and easily distinguishable alternative hypotheses:

$$\text{lag}_1 \, X^*_{qj} = \sum w^1_t X^*_{q+t,j} \qquad j=8, 9, 10 \qquad (4.19.7)$$

for which a mode of 1·5 was used;

and

$$\text{lag}_4 X^*_{qj} = \sum_t w^4_t X^*_{q+t,j} \qquad j=8, 9, 10 \qquad (4.19.8)$$

for which the mode was 4·5.

The weights w^1 imply a maximum rate of adjustment in the moving-quarter following that in which the price changed, that is, the quarter in which credit gas and electricity consumers receive their first bill at the new price, and a greater rate of adjustment in the quarter of the change than given by the weights w^0. Although the complete adjustment takes three years, 98% of it is in the first two.

Weights w^4 imply a maximum rate of adjustment one year after a price change, and no adjustment in the quarter of the change. On average, adjustment is later by over two quarters compared with that given by weights w^1. The implied hypothesis is that habits refer to particular seasons and are not changed much until each season has been experienced at the new price.

The hypothesis which leads us to include lagged values of $(\bar{E}/\bar{\pi})$ is as follows: either a change in current income, I, (excluding cyclical changes) produces an immediate change in income expectations and therefore I^*, as defined in 4.17 above, and this leads, because of the strength of habits, to only a gradual U effect; or a change in I produces a gradual revision of income expectations and I^* and each stage of the revision to an instantaneous U effect; or more plausibly some combination of both.

One variable was calculated:

$$\left. \begin{array}{c} \text{lag}_0 \, \ln (\bar{E}/\bar{\pi})_q = \sum_{t=-15}^{0} w^0_t \ln (\bar{E}/\bar{\pi})_{q+t} \\[2mm] w^0_t = \frac{1}{16} \qquad t = -15, ..., 0 \end{array} \right\} \qquad (4.19.9)$$

where

A four-year period is used rather than a shorter one because our hypothesis is that cyclical changes in current income did not lead to changes in utilization, and four years was roughly the period of the cycles. The U effect is unlikely to have been spread over a longer period. A simple linear adjustment path is postulated because there is no information from which to derive a more elaborate hypothesis. As the series for $\bar{E}/\bar{\pi}$ is smoother than the price series, there is no prospect of being able to discriminate between alternative hypotheses on the basis of regression results.

Total consumption expenditure, the price of solid fuel and the index of all other prices were subject to regular seasonal variations. We hypothesise that utilization did not adjust to these regular changes and therefore use seasonally adjusted values—$\bar{E}$, $\bar{P}^c$, $\bar{\pi}$. The quarterly values of $\bar{E}/\bar{\pi}$ were obtained by interpolating linearly between mid-year values of E/π. For deflating the price variables, $\bar{\pi}$ in moving-quarter q is defined as

$$\bar{\pi}_q \equiv \tfrac{1}{4} \sum_{q=t-3}^{t} \pi_t \qquad (4.19.10)$$

$\bar{P}_q^c$ in moving-quarter q is defined as

$$\bar{P}_q^c \equiv \tfrac{1}{4} \sum_{q=t-3}^{t} P_t^c \qquad (4.19.11)$$

These forms have been used in order that future prices do not appear in the utilization functions. They imply a slightly slower response to changes in all other prices and the price of solid fuel than to changes in the prices of electricity and gas. It seems plausible anyway that changes in fuel prices would have a more immediate effect on fuel demand than changes in other prices, and that stockholding would tend to delay the effect of changes in the price of solid fuel.

As the lagged values are determinants of ownership, their inclusion in the utilization functions increases the likelihood that the coefficients in the functions will take up the explanation of Z-U effects occurring over time as well as those arising from inter-Area differences in ownership, and therefore makes assumption 1.5.2, justifying the use of pooled data, more nearly valid.

4.20 The natural variables—L, W

The technical literature contains the results of a large amount of research into the relationship between demand and natural variables,[11] most of it at a lower level of aggregation than the present study—usually hourly or instantaneous demand is the dependent variable, and some studies are of small groups of consumers.

To summarize the main results of relevance to this study:

(1) Space-heating demand depends on the effective room temperature as determined by the external air temperature, the rate of heat loss of the building—which in turn depends on the wind speed and form of construction—, humidity and rainfall. Room temperature depends

(11) See especially Davies (1958) for a comprehensive analysis of the natural determinants of the total instantaneous demand from all users; Electricity Council *Utilization Research Reports* nos. 55 to 58 on space-heating, 66 on water-heating and 65 on lighting.

on external air temperatures over the previous 24 hours, the weights declining for successively earlier hours. At any given time of day there is a non-linear response to a change in room temperature.

(2) Water-heating demand depends on the water supply temperature which bears a lagged and damped relationship to air temperature, the lag being of the order of one month.

(3) Lighting demand during a day depends on the amount of natural illumination as determined by the hours of daylight, cloud cover, and visibility.

(4) For all three classes, the responsiveness to changes, in kilowatts per consumer, depends on the time of day.

Although meteorological data are available in abundance, the form is such that for many variables it is impracticable to obtain the required monthly series of means for all consumers in each Area. We have therefore restricted our attention to a series for air temperature to be used in explaining both space-heating and water-heating demand, and a series for hours of daylight. Preliminary inspection of data on other variables showed that, given the large element of guesswork in weighting different recordings to derive monthly and Area means and the relatively low explanatory power (when analysing quarterly series) of variables other than the two selected, no improvement in the regression results could be expected from using them.

The temperature variable W is a simple average of the temperatures at 09.00 and 21.00 hours Greenwich Mean Time. Two have been used so that some account is taken of temperatures close to both the morning and evening periods of high space-heating demand. The error due to using temperatures recorded at varying clock times should be small as the errors in the two temperatures are in opposite directions.

In Areas other than London and Merseyside & North Wales no one recording station could be regarded as representative (even after correction for altitude) of the whole Area.[12] Areas were therefore divided into up to five isothermal zones, for example coastal and inland, and data collected for each. Temperatures were adjusted to the estimated mean altitude of consumers in a zone where this differed from the altitude of the recording station. Area means for each month were obtained using the estimated proportion of consumers in each zone as weights.

With a given demand for heat at a particular time of day, there is some temperature above which a consumer would use no electric space-heating. Below this temperature, heaters would be brought into use until at some temperature the total installed load is in use. Aggregating over consumers, a complete hourly demand curve would probably then be sigmoid. Aggregating

(12) See Lyness & Badger (1970) section 4.1.

over time to obtain monthly demand, it is very unlikely that the flattening of the curve at low temperatures would occur within the range of temperatures experienced. Our hypothesis is therefore that as temperature rose within the observed range, demand fell at a decreasing rate until at some temperature τ^u there was virtually no temperature response: the function in (3.5.2) is defined as

$$X_3 \equiv \ln W^u \qquad (4.20.1)$$

where $W_m^u = \min (W_m, \hat{\tau}^u)$, that is, the lower of the calculated monthly value and some temperature to be estimated in the regression analysis.[13]

It is implied that inter-Area and temporal variations have the same effect on demand, that is, the desired room temperature is independent of the average air temperature experienced. The similar hypothesis is made for lighting that the desired level of illumination at any given clock time is independent of latitude.

Domestic load curves and evidence on sleeping habits[14] suggest that a household had a positive demand for illumination from on average soon after 07.00 hours, and in the evening until about 23.00 hours. As the astronomical definitions of sunrise and sunset are used, solar illumination would not be great enough to meet the demand until somewhat more than half an hour after sunrise. Hence variations in the time of sunset would always tend to produce variations in the demand for electric lighting, but only variations in the time of sunrise after 06.30 would do so to any significant extent. The daylight variable L is therefore defined as the minutes of daylight between sunrise or 06.30 clock time, whichever is the later, and sunset. In order to obtain the Area series, the mean latitude of consumers in each Area was estimated to the nearest half degree.

As the domestic load curves suggest that the aggregate demand for illumination rose fairly steadily from virtually zero at 06.00 to a peak at 08.00, successive unit decreases in the period of daylight before 08.00 might be expected to produce successively larger increments in the demand for electric lighting. Since a similar argument applies to the end of the day, the relationship between hours of daylight and lighting demand should be nonlinear. As the semi-log form appears likely to be a good approximation over the observed range, the function in (3.5.3) is specified as

$$X_4 \equiv \ln (L/\bar{L}) \qquad (4.20.2)$$

where $\bar{L}$ is an average, introduced for convenience in estimation.[15]

---

(13) The threshold has had to be introduced into the calculation after aggregating over consumers. Hence the mean X_3 differs from that defined by (3.5.9) in some quarters.

(14) See Kleitman (1963).

(15) See 6.2 where L is defined.

The main effects of using only these two natural variables, X_3 and X_4, are (1) some loss of explanatory power as differences in responsiveness at weekends, holidays and different times of day have been ignored; and (2) a change in the absolute values of their coefficients. Omitting the mean values of other variables will tend to raise them; ignoring variability within months, and using air temperature in the water-heating utilization function, U^2, will tend to lower them.

Variations in U^2 due to variations in temperature between winter months will probably be poorly explained: all-the-year-round users of electric water-heaters may be assumed to have had the same form of relationship between U^2 and temperature as that between space-heating utilization U^1 and temperature postulated above, so that in the winter aggregate U^2 was a decreasing function of temperature. But there were many households which only used their electric water-heating during the summer. If these were a sufficiently high proportion, there would have been a middle range of temperatures in which aggregate U^2 was an increasing function of temperature—and a higher range in which the curve flattened out and then fell again. If this middle range was fairly wide and includes enough observations, the temperature coefficient will be positive.

Errors in the variables, for example temperatures too high on average in a particular Area or displaying the wrong seasonal variation because a recording station is unrepresentative, and the omission of other variables moderately highly correlated with the two included, will tend to produce a seasonal and Area pattern in the regression residuals.

CHAPTER 5

ESTIMATION AND TESTING: METHODS

5.1 Introduction

We turn now to a consideration of how to estimate and test the model. Our objectives are:

(1) to estimate the parameters by a method based on assumptions which are as nearly as possible valid for the model defined in Chapter 3 and the set of 624 pooled observations with the properties described in Chapter 4;

(2) to measure the reliability of the estimated coefficients and the goodness of fit of the equation; to test the specification of the equation with respect to the variables included, the constancy of the parameters and the properties of the error term—in particular, to examine whether there is any Area or seasonal pattern in the residuals or serial correlation; and to make predictions;

(3) as a prerequisite of (2), to preserve arithmetic accuracy during the computations.

5.2 Notation

To simplify the exposition in this chapter, the equation to be estimated (4.4.10), is rewritten as

$$\mathbf{Y} = \mathbf{X}\boldsymbol{\beta} + \boldsymbol{\varepsilon} \tag{5.2.1}$$

$\mathbf{Y}$ is an $(m \times 1)$ vector: $\mathbf{Y} = \| Y_i \|$
where $Y_i = Y_{qB}^u$
    m is the number of observations (624)
    B the Area $(B = 1, 2, ..., 12)$
    q the moving-quarter $(q = 1, 2, ..., 52)$
and observations are ordered by quarter for each Area so that

$$i = 52\,(B-1) + q$$

$\mathbf{X}$ is an $(m \times n)$ matrix: $\mathbf{X} = \| x_{ik} \|$

where $x_{ik} = 1 \qquad k = 1$
$$= Z^a_{qB} X_{qB,j} \qquad \left.\begin{array}{l} a = 1, 2, 3, 4; j = 2, 3, \ldots, 11; \\ k = 2, 3, \ldots, n \end{array}\right\} \qquad (5.2.2)$$

n is the number of variables of this form and variables are ordered by j for each class of appliance, a, so that

$$k = 4(j-1) + a - 3.$$

The kth column of $\mathbf{X}$ is written $\mathbf{x}_k$.[1]

β is an $(n \times 1)$ vector: $\beta = \| \beta_k \|$
where $\beta_k = \beta_{j,a}$.

ε is an $(m \times 1)$ vector: $\varepsilon = \| \varepsilon_i \|$.

5.3 The method of estimation

By assumption 3.10.1,

$$\begin{aligned} E\{\varepsilon\} &= \mathbf{0} \\ E\{\varepsilon.\varepsilon'\} &= \sigma^2 \mathbf{V} \end{aligned}$$

where $\mathbf{V}$ is $(m \times m)$ with submatrices $\mathbf{V}_B$ as defined in 4.5 on the leading diagonal and zeros elsewhere.

We first obtain some results making the following further assumptions and then examine the effect of their invalidity in 5.10 below.

Assumption 5.3.1

The elements of ε are distributed Normally.

Assumption 5.3.2

There are no errors in $\mathbf{V}$ as calculated.

Assumption 5.3.3

$\mathbf{X}$ has full column rank, n.

Assumption 5.3.4

There are no errors in $\mathbf{X}$ as calculated.

Given these assumptions, the optimal estimator of β is obtained by the method of Generalized Least Squares.[2] This is

$$\mathbf{b} = (\mathbf{X}'\mathbf{V}^{-1}\mathbf{X})\mathbf{X}'\mathbf{V}^{-1}\mathbf{Y} \qquad (5.3.1)$$

In particular, since $\mathbf{V}$ is not the identity matrix, Generalized Least Squares is superior to Ordinary Least Squares.

(1) In Chapter 6 other definitions of $\mathbf{x}_k$ as a function of the Z^a and X_j variables are used. The results of this chapter are unaffected.
(2) See, e.g., Malinvaud (1966) part 2.

5.4 The method of computation

Given that the model specifies $\mathbf{X}$ in such a way that some variables are highly correlated, the standard methods of calculating Ordinary Least Squares and Generalized Least Squares estimators on electronic computers are unsatisfactory: inversion of $\mathbf{X}'\mathbf{V}^{-1}\mathbf{X}$ by, for example, Gauss-Jordan elimination would be likely to produce arithmetically inaccurate results,[3] and also, on the computers used for most of our regression analysis, posed problems on account of the amount of store used.

Instead orthonormalization by means of the Cholesky decomposition has been used.

Since the error variance-covariance matrix is positive definite, it may be written

$$\mathbf{V} = \mathbf{P}.\mathbf{P}' \tag{5.4.1}$$

where $\mathbf{P}$ is lower triangular $(m \times m)$.[4]

(5.2.1) is to be estimated by

$$\mathbf{Y} = \mathbf{X}\mathbf{b} + \mathbf{e} \tag{5.4.2}$$

where $\mathbf{e}$ is an $(m \times 1)$ vector of residuals: $\mathbf{e} = \| e_i \|$

Writing $\mathbf{R} = \mathbf{P}^{-1}$ (5.2.1) becomes, on premultiplying by $\mathbf{R}$,

$$\mathbf{RY} = \mathbf{RX}\beta + \mathbf{R}\varepsilon \tag{5.4.3}$$

Since $\quad\quad E\{(\mathbf{R}\varepsilon).(\mathbf{R}\varepsilon)'\} = \sigma^2\mathbf{I}_m$

where $\mathbf{I}_m$ is the $(m \times m)$ identity matrix, Ordinary Least Squares may be applied to (5.4.2) to obtain

$$\mathbf{b} = [(\mathbf{RX})'(\mathbf{RX})]^{-1}(\mathbf{RX})'(\mathbf{RY}) \tag{5.4.4}$$

Using the same decomposition again, since the matrix in square brackets is positive definite, we may write

$$[(\mathbf{RX})'(\mathbf{RX})] = \mathbf{S}.\mathbf{S}' \tag{5.4.5}$$

where $\mathbf{S}$ is lower triangular $(n \times n)$.

Suppose that Ordinary Least Squares is used to obtain estimators $\mathbf{b}^\circ$ from

$$\mathbf{RY} = \mathbf{RXT}'\mathbf{b}^\circ + \mathbf{e}^\circ \tag{5.4.6}$$

where $\mathbf{T} = \mathbf{S}^{-1}$. Then

$$\mathbf{b}^\circ = \mathbf{T}(\mathbf{RX})'(\mathbf{RY}) \tag{5.4.7}$$

(3) See Longley (1967). Since the publication of Longley's article there has been a rapidly growing literature in which the seriousness of the inaccuracy has been confirmed—see Wampler (1970).

(4) See, e.g., Plackett (1960) pp. 2-4 for a proof and the methods of computation given in the rest of this section.

Comparing (5.4.4) and (5.4.7),

$$\mathbf{b} = \mathbf{T}'\mathbf{b}^\circ \qquad\qquad (5.4.8)$$

and

$$\mathbf{e}^\circ = \mathbf{e} \qquad\qquad (5.4.9)$$

Hence $\mathbf{b}$ may be obtained by first computing the coefficients in the ortho-normal form by (5.4.7) and then using (5.4.8), avoiding the inversion of both $\mathbf{V}$ and $\mathbf{X}'\mathbf{V}^{-1}\mathbf{X}$.

The elements of $\mathbf{S}$ are obtained as follows:
let the typical element of $\mathbf{X}'\mathbf{V}^{-1}\mathbf{X}$ be w_{ij};
then

$$s_{11} = \sqrt{w_{11}}$$

$$s_{i1} = \frac{w_{1i}}{s_{11}} \qquad i = 2, \ldots, n$$

$$s_{ii} = \sqrt{\left(w_{ii} - \sum_{j=1}^{i-1} s_{ij}^2\right)} \qquad i = 2, \ldots, n$$

$$s_{ki} = \frac{w_{ik} - \sum_{j=1}^{i-1} s_{ij}s_{kj}}{s_{ii}} \qquad k = i+1, \ldots, n; \; i = 2, \ldots, n-1$$

$$= 0 \qquad k < i$$

Hence the elements of $\mathbf{T}$ are

$$t_{11} = \frac{1}{s_{11}}$$

$$\left.\begin{array}{l} t_{ii} = \dfrac{1}{s_{ii}} \\[2mm] t_{ij} = \dfrac{-\sum_{k=j}^{i-1} s_{ik}t_{kj}}{s_{ii}} \end{array}\right\} i = 2, \ldots, n; \; j = 1, \ldots, i-1$$

$$t_{ij} = 0 \qquad i < j$$

The elements of $\mathbf{R}$ are obtained by the same method except that some simplification is possible since elements of $\mathbf{V}$ more than one away from the leading diagonal are zero.

Tests with ill-conditioned matrices confirmed that this method achieves an improvement in arithmetic accuracy compared with matrix inversion by Gauss-Jordan elimination as implemented in the standard programs on the machines used. It has the further advantages of being economical in storage, allowing the same algorithm to be used for $\mathbf{R}$ and $\mathbf{T}$, and producing the co-efficients $\mathbf{b}^\circ$ required for tests on groups of the coefficients $\mathbf{b}$. Whilst the accuracy was not as great as could have been achieved by using, say, one of the Gram-Schmidt orthogonalization procedures,[5] it reached a satisfactory level with our electricity data: no significant figures were lost compared with the original data.

5.5 Test statistics, measures of reliability and goodness of fit

The following results may be proved for Generalized Least Squares[6]:

$$E\{\mathbf{b}\} = \beta \tag{5.5.1}$$

$$E\{(\mathbf{b}-\beta).(\mathbf{b}-\beta)'\} = \sigma^2(\mathbf{X}'\mathbf{V}^{-1}\mathbf{X})^{-1} \tag{5.5.2}$$

$$E\{(\mathbf{Re})'.(\mathbf{Re})\} = (m-n)\sigma^2 \tag{5.5.3}$$

Hence
$$\hat{\sigma}^2 \equiv \frac{\mathbf{e}'\mathbf{V}^{-1}\mathbf{e}}{m-n} \tag{5.5.4}$$

is an unbiased estimator of σ^2, which we shall call the *error variance*;[7] and

$$\hat{\sigma}_k^2 \equiv \hat{\sigma}^2 \sum_{j=k}^{n} t_{jk}^2 \tag{5.5.5}$$

is an unbiased estimator of the variance of b_k.

As $m = 624$ and $n < 27$ in all the analyses in Chapter 6, the 95% confidence interval for β_k is

$$b_k \pm 1\cdot 96\, \hat{\sigma}_k \tag{5.5.6}$$

For testing hypotheses about groups of coefficients, test statistics may be derived as follows:[8]

let
$$\bar{y} = \frac{\mathbf{1}'\mathbf{V}^{-1}\mathbf{Y}}{\mathbf{1}'\mathbf{V}^{-1}\mathbf{1}} \tag{5.5.7}$$

$\bar{y}_i$ be the typical element of the vector obtained by subtracting $\bar{y}$ from each element of $\mathbf{RY}$,

and e_i^* be the typical element of $\mathbf{Re}$.

Then, from (5.4.6),

$$\sum_{i=1}^{m} \bar{y}_i^2 = \sum_{k=2}^{n}(b_k^\circ)^2 + \sum_{i=1}^{m}(e_i^*)^2 \tag{5.5.8}$$

Since
$$E\{(\mathbf{b}^\circ - E\{\mathbf{b}^\circ\}).(\mathbf{b}^\circ - E\{\mathbf{b}^\circ\})'\} = \sigma^2 \mathbf{I}_n \tag{5.5.9}$$

and
$$E\{(\mathbf{Re}).(\mathbf{b}^\circ - E\{\mathbf{b}^\circ\})\} = 0 \tag{5.5.10}$$

(5) See Longley (1967) and Wampler (1970).
(6) The proofs are simple generalizations of the usual textbook results for Ordinary Least Squares. See, e.g., Theil (1971) chapter 6.
(7) It is not the variance of any one error since $v_{ii} \neq 1$ any i.
(8) Johnston (1963) chapter 4 gives the Ordinary Least Squares counterparts.

$\dfrac{(b_k^o - E\{b_k^o\})^2}{\sigma^2}$ $k = 2, \ldots, n$ are independent χ^2-variates with 1 degree of freedom and $\sum(e_i^*)^2/\sigma^2$ an independent χ^2-variate with $m-n$ degrees of freedom.

Since $$\mathbf{b}^o = \mathbf{S}'\mathbf{b}$$

and $\mathbf{S}'$ is upper triangular, a necessary and sufficient condition for

$$b_k = 0 \qquad k = p+1, \ldots, n;\ 0 < p < n$$

is $$b_k^o = 0$$

Hence the hypothesis $H_0: \beta_k = 0\ (k = p+1, \ldots, n)$ may be tested using the test statistic

$$\frac{\sum\limits_{k=p+1}^{n} (b_k^o)^2/(n-p)}{\sum(e_i^*)^2/(m-n)} \tag{5.5.11}$$

which is distributed as $F(n-p,\ m-n)$. We shall denote this $F\{b_{p+1}^o, \ldots, b_n^o\}$ and in general use $F\{\ldots\}$ to denote the F-test statistic for the test of the hypothesis that all the coefficients whose estimators are listed inside the brackets are zero. Similarly $t\{\ldots\}$ will denote the t-test statistic.

For $p = 1$, the test is equivalent to a test of the hypothesis that the coefficient of determination R^2, defined below, does not differ significantly from zero.

For $p = n-1$, (5.5.11) is equal to

$$\frac{b_k^2}{\hat{\sigma}_k^2} \tag{5.5.12}$$

and distributed as $F(1,\ m-n)$. This may be used for a test of the hypothesis $H_0: \beta_k = 0$ for any k.

To measure the goodness of fit of the equation, we calculate

$$R^2 = 1 - \frac{\sum(e_i^*)^2}{\sum \bar{y}_i^2} \tag{5.5.13}$$

The usual property of the coefficient of determination, $0 \leqslant R^2 \leqslant 1$, then carries over to estimation by Generalized Least Squares.

The same definition of R^2 has been used where an equation contains no constant.[9] In this case it could be negative.

Correlation between variables is measured by zero-order correlation efficients defined as

(9) See Theil (1971) p. 178 note 4.

$$r(\mathbf{x}_j, \mathbf{x}_k) = \frac{\sum_i (x_{ij} - \bar{x}_j)(x_{ik} - \bar{x}_k)}{\sqrt{(\sum_i (x_{ij} - \bar{x}_j)^2 \sum_i (x_{ik} - \bar{x}_k)^2)}} \qquad (5.5.14)$$

where $\bar{x}_k$ is the simple mean of the 624 moving-quarter figures x_{ik}.

5.6 Analysis of the pattern of residuals

An analysis of the pattern of residuals is undertaken to throw some light on the extent of specification error, particularly that leading to loss of efficiency and underestimation of standard errors. We look for three kinds of pattern:

(1) Area and seasonal variation;
(2) serial correlation, after any seasonal variation has been eliminated;
(3) correlation between the absolute size of the residuals and some other variable or variables in the model, suggesting heteroscedasticity of the elements of $\mathbf{R}\varepsilon$.

The hypothesis that the errors had the same constant absolute seasonal pattern in each Area and varied in their mean level between Areas may be expressed by writing

$$\mathbf{R}\varepsilon = \mathbf{D}\gamma + \xi \qquad (5.6.1)$$

where γ is a (15×1) vector of coefficients,
  ξ an $(m \times 1)$ vector of random errors with zero expectations and constant variance σ_ξ^2
and $\mathbf{D}$ an $(m \times 15)$ matrix of dummy variables $\mathbf{D}_k$.

$$\mathbf{D}_1 = 1$$

$\mathbf{D}_2$, $\mathbf{D}_3$, $\mathbf{D}_4$ take the value 1 only in moving-quarters II, III, IV respectively and 0 otherwise; $\mathbf{D}_5$ to $\mathbf{D}_{15}$ take the value 1 only in Areas 1, 2, 3, 5, ..., 12 respectively and 0 otherwise.

An estimator $\mathbf{c}$ of γ is obtained by applying Ordinary Least Squares to

$$\mathbf{R}e = \mathbf{D}\mathbf{c} + \mathbf{z} \qquad (5.6.2)$$

where $\mathbf{z}$ is an $(m \times 1)$ vector of residuals: $\mathbf{z} = \| z_i \|$.

Because the dummy variables $\mathbf{D}$ and the variables $\mathbf{X}$ are not orthogonal (after removal of means), $\mathbf{c}$ is not in general the same as the estimator of γ obtained by estimating β and γ from one regression. But since our purpose is to examine the pattern in what is left *after the variables* $\mathbf{X}$ *have explained as much as possible*, this does not matter.

Use of $\mathbf{R}e$ instead of the unknown $\mathbf{R}\varepsilon$ means that $\mathbf{c}$ does not possess the usual optimal properties of Ordinary Least Squares estimators, but it should still be satisfactory to base conclusions on standard errors and F-statistics calculated by the standard formulae when comparing the residuals from regressions differing only in respect of a few of the explanatory variables included.

As it is useful to have F-statistics for the groups of seasonal and Area coefficients, $\mathbf{c}$ is computed, by the same procedure as given above, from the coefficients $\mathbf{c}^o$ in the orthonormalized form

$$\mathbf{Re} = \mathbf{DF'c}^o + \mathbf{z} \qquad (5.6.3)$$

where $\mathbf{F}$ is the (15×15) lower triangular matrix satisfying

$$(\mathbf{D'D})^{-1} = \mathbf{F'F} \qquad (5.6.4)$$

Writing $\hat{\sigma}_\xi^2 \equiv \sum z_j^2 / (m - 15)$, the F-statistic for the group of Area coefficients is

$$\frac{\sum\limits_{k=5}^{15} (c_k^o)^2 / 11}{\hat{\sigma}_\xi^2} \qquad (5.6.5)$$

Since the groups of seasonal and Area dummies are orthogonal, with means taken out,

$$\frac{\sum\limits_{k=2}^{4} (c_k^o)^2 / 3}{\hat{\sigma}_\xi^2} \qquad (5.6.6)$$

may be used for a test for the presence of seasonal pattern alone.

5.7 Testing for serial correlation

The residuals z_i could display serial correlation:

  (1) for the reasons applicable to a regression on a single time-series—in particular, omission of variables, non-linearity in the variables included, misspecification of lags and serial correlation of errors in the variables, especially due to the use of interpolated values;

  (2) because of inter-Area differences in parameters—given that many of the included variables changed fairly gradually, these differences could lead to strong positive serial correlation, varying between Areas;

  (3) because not all the seasonal pattern has been removed by the dummies—if there are inter-Area variations in the pattern, the residuals z_i will display less seasonal variation in some Areas than the e_i^*, and in others a reversed (and possibly more marked) seasonal pattern.

It is possible to attempt an assessment of the relative importance of these causes by constructing a single average measure of the serial correlation in all twelve Area time-series, an average from which the influence of any seasonal pattern has been removed, and relative measures of the correlation in each of the twelve series separately.

The hypothesis tested using the Durbin-Watson d statistic may be expressed as follows:[10]

let
$$\xi_i = \rho \xi_{i-1} + \eta_i \qquad i = \ldots -1, 0, 1, \ldots \qquad (5.7.1)$$

where η_i is a random error distributed Normally with mean zero and constant variance and independently of ξ_{i-1}, ξ_{i-2}, ..., η_{i-1}, η_{i-2}, ... Then the hypothesis under test is $H_0: \rho = 0$.

The use of Durbin-Watson d is inappropriate when the observations do not form a single time-series, but a test of the same hypothesis may be devised: suppose that the parameter ρ takes the same value in each time-series. Then it may be estimated from

$$z_i = \hat{\rho} z_{i-1} + h_i \qquad i = 52(B-1) + q; \; B = 1, 2, \ldots, 12; \; q = 2, 3, \ldots, 52 \quad (5.7.2)$$

Applying Ordinary Least Squares,

$$\hat{\rho} = \frac{\sum\limits_{2}^{m} z_i z_{i-1}}{\sum\limits_{2}^{m} z_{i-1}^2} \qquad (5.7.3)$$

As z_{i-1} is a lagged dependent variable, $\hat{\rho}$ is a biased but consistent estimator of ρ. Hurwicz[11] has shown that it is biased towards zero and that for small values of ρ,

$$\lim_{|\rho| \to 0} \frac{E(\hat{\rho})}{\rho} = 0 \cdot 9804 \text{ for } m = 100$$

$$= 0 \cdot 9960 \text{ for } m = 500.$$

He has also shown that for small m (3 and 4), the relative bias $E(\hat{\rho})/\rho \to 1$ monotonically as $|\rho| \to 1$. Assuming that this is true for higher m—Hurwicz could only conjecture that this is so—the relative bias should be very close to unity for any ρ with $m = 624$. Similarly the bias in the estimated variance

$$\hat{\sigma}_\rho^2 = \frac{\sum h_i^2 / (m-13)}{\sum z_{i-1}^2} \qquad (5.7.4)$$

should be small. Hence given $H_0: \rho = 0$, the distribution of $\hat{\rho}/\hat{\sigma}_\rho$ should closely approximate that of Student's t with $m-13$ degrees of freedom. This statistic is used to test for the presence of serial correlation in the twelve time-series together, including that due to seasonality.

(10) See Durbin & Watson (1950) p. 423.
(11) Hurwicz (1950).

To remove the effects of any regular seasonal pattern, we estimate by Ordinary Least Squares the coefficient $\hat{\rho}_a$ in

$$z_i^a = \hat{\rho}_a z_{i-1}^a + h_i^a \qquad i = B+k; \; B = 1, 2, \ldots, 12; \; k = 2, \ldots, 13 \tag{5.7.5}$$

where z_{B+k}^a is the sum of the four residuals in moving-quarters $4(k-1)+1$, $\ldots, 4(k-1)+4$ in Area B. By the same argument as above, the ratio of $\hat{\rho}_a$ to its estimated standard error may then be used to provide a test of H_0: $\rho_a = 0$, where ρ_a is the coefficient of autocorrelation between successive annual sums of errors defined similarly to ρ in (5.7.1).

To measure inter-Area differences in the serial correlation, we calculate

$$\frac{d_B = \sum_{q=2}^{52} (z_{52(B-1)+q} - z_{52(B-1)+q-1})^2}{\sum_{q=2}^{52} (z_{52(B-1)+q})^2} \qquad B = 1, 2, \ldots, 12 \tag{5.7.6}$$

The residuals z_i are obtained from a regression in which Area constants are fitted but do not otherwise satisfy Durbin & Watson's assumptions, so that the critical values given by them for d can only be used as approximate guides to the significance of the absolute level of correlation in any series. But the figures for d_B should provide a satisfactory indication of relative levels in the different series.

5.8 Testing for heteroscedasticity

Systematic variation in the absolute size of the residuals **Re** may arise because of variations in the Area or seasonal means due to incorrect specification of the model as described in 5.6 above, or because of heteroscedasticity of the corresponding errors, **Rε**. The test of the hypothesis of homoscedasticity is therefore based on the residuals **z** as these are not subject to variation from the first cause.

The most obvious alternative hypothesis to homoscedasticity is that the error variance is related to the expected value of the dependent variable, given the values of the independent variables, a specific hypothesis often used elsewhere being that it is proportional to the square of the dependent variable.[12] We shall suppose that

$$E\{\xi_i^2\} = \sigma^2 [E\{y_i\}]^{2\psi} \qquad \text{all } i \tag{5.8.1}$$

where y_i is the ith element of **RY**. Then a test is required of

$$H_0: \psi = 0$$

against H_1: $\psi \neq 0$, or the more specific alternative $\psi = 1$. An approximate test is provided by estimating $\hat{\psi}$, by Ordinary Least Squares, in

(12) Compare the discussion of assumption 3.10.1 in 3.10. See also Cramer (1969) p. 86.

$$\ln (z_i^2) = k + \hat{\psi} \ln (y_i^2) + d_i \qquad i = 1, 2, \ldots, m \qquad (5.8.2)$$

where k is a constant and the d_i are residuals, and using that, given H_0, $\hat{\psi}$ divided by its estimated standard error is approximately distributed as Student's t with $m - 2$ degrees of freedom.

If ξ_i is Normally distributed as assumed above, the error δ_i in

$$\ln (\xi_i^2) = \kappa + \psi \ln (E\{y_i\})^2 + \delta_i \qquad (5.8.3)$$

is not Normally distributed. This and the replacement of the unknown ξ_i by z_i mean that $\hat{\psi}$ is not exactly distributed as t with $m - 2$ degrees of freedom but the approximation should be good given the large number of degrees of freedom.

y_i has been used instead of $E\{y_i\}$ to free the test from dependence on the assumption that the true relationship between $\mathbf{Y}$ and $\mathbf{X}$ is linear. This is achieved at the expense of introducing a downward bias[13] in $\hat{\psi}$ if the true relationship is linear, due to the error in the variable. But for high values of R^2 (as obtained), this is small, and a worthwhile improvement in the estimation of ψ may be expected if non-linearity is present.

A possible difficulty in using (5.8.2) is that, as the z_i are distributed around zero, a zero may occur. The program was therefore designed to substitute $0 \cdot 0001$ for any residual z_i which was smaller in absolute value than this. Actually no residuals as small as this occurred in any of the regressions performed.

5.9 Prediction

The availability of data for the South of Scotland Electricity Board Area on both the dependent and independent variables enables us to make predictions from the regression results and to check them. Whilst nothing is gained or lost by omitting the South of Scotland observations from the set of observations on which the regression analysis is performed, it is convenient to omit them because the time-series are shorter and some are subject to wider margins of error than the corresponding series for Areas in England and Wales.

The predictive test is a valuable supplement to the tests already described as it is a severe test of the specification of the functional form: for many quarters the values of the dependent variable and the variables measuring temperature, hours of daylight, the ownership of off-peak appliances, the price of electricity and the price of gas all lie outside the ranges of values in Areas in England and Wales.

If the true value of the dependent variable in the South of Scotland Area in moving-quarter q is

$$Y_{q13}^u = \mathbf{x}_{q13}' \, \boldsymbol{\beta} + \varepsilon_{q13} \qquad (5.9.1)$$

(13) See, e.g., Johnston (1963) p. 150.

where x'_{q13} is the row vector of observations on the independent variables, then the best point predictor of Y^u_{q13} is

$$\hat{Y}_{q13} \equiv x'_{q13} b \qquad (5.9.2)$$

Assuming that ε_{q13} is independent of all elements of ε and that V_B is the same for $B=13$ as for $B=1, 2, \ldots, 12$, so that

$$E\{\varepsilon^2_{q13}\} = \sigma^2 v_{qq} \qquad (5.9.3)$$

the variance of the predictor is best estimated by

$$\hat{\sigma}^2 [(x'_{q13} T')(x'_{q13} T') + v_{qq}] \qquad (5.9.4)$$

5.10 Invalidity of the assumptions

Returning now to the assumptions made in 5.3 and 3.10, we need to examine what effect their invalidity has on the properties of the estimators and the critical values of the test statistics for a given Type I error.

If all the assumptions are valid, then for $m=624$ observations and all the values of n, the number of variables, used in Chapter 6, the critical F-value for a one-tailed test of a hypothesis about a single coefficient at the 5% level of significance is 3·84, and the critical t-value for a two-tailed test at the 5% level 1·96. Hereafter "significant" will always mean significant at the 5% level, and all confidence intervals will be 95% ones.

Invalidity of *assumption 5.3.1* on Normality of the errors is much less important than invalidity of the other assumptions. The assumption is only required to justify the tests, and they are not very sensitive to departures from Normality.[14]

Given assumption 3.10.1 on homoscedasticity of the daily errors, any inaccuracy in V must be very small as a proportion of $\sigma^2 v_{qq}$ for any q: even if *assumption 5.3.2* is not strictly valid, use of V for Generalized Least Squares estimation (in particular use of the positive covariance terms) will produce a worthwhile gain in efficiency compared with Ordinary Least Squares.

Assumption 5.3.3 on the rank of X is valid for all the sets of variables used. Where it is possible to use prior information to constrain parameters so that highly correlated variables may be omitted, this is done (as described in Chapter 6). But where the arguments of the previous chapters lead to the inclusion of two or more highly correlated variables about the coefficients of which there is no prior information, they are left in the equation. The lowness of their individual F-values then provides a warning of their unreliability; the F-test on the group of coefficients provides a satisfactory indication of the explanatory power of the variables as a group.

Invalidity of *assumption 5.3.4*—no errors in variables—leads to biased

(14) See, e.g., Malinvaud (1966) chapter 3 paragraph 7 and chapter 8 paragraph 4.

estimators. The magnitude of the errors in X has been considered in Chapter 4. As the information on the errors does not lend itself to integration within the formal specification of the model, the procedure adopted in Chapter 6 is to give the estimates without any adjustment and to indicate the likely bias when commenting on them.

The use of interpolated values also has implications for the number of degrees of freedom attached to test statistics. Given the form of the variables x_k in (5.2.2), no exact rule can be formulated, but bounds can be placed on the critical value of a test statistic. Suppose that there were $p-1$ variables in X which were functions only of variables in the calculation of which interpolation was used; and that the remaining $n-p$ variables (excluding the constant) were functions only of variables orthogonal to those in the first set after removing means and in the calculation of which no interpolation was used. Further suppose that, as for certain appliance ownership variables, all but 36 of the 624 values were interpolated if interpolation was used at all. Then the p coefficients could be estimated by a separate regression; since there would be no gain from using the interpolated values, the degrees of freedom would be $36-p$. Given that for most variables there were more than 36 uninterpolated values, and for some 624, this result suggests that, given assumption 3.10.1, a value of an F-test statistic over about 4·2 could be regarded as refuting the hypothesis under test but a value between 3·84 and 4·2 must be regarded as inconclusive.

The other basic assumption is *3.10.1* on the independence and homoscedasticity of the errors. If serial correlation is shown to be present, the variances of b, c and $\hat{\psi}$ will be underestimated thereby increasing the probability of rejecting wrongly the hypotheses that a variable in X has no explanatory power, that there is no Area or seasonal pattern in the residuals and that there is homoscedasticity.

The presence of serial correlation or heteroscedasticity implies a loss of efficiency. Computational difficulties arising from the form of V preclude the use of more efficient methods[15] with moving-quarter data. From the low values of $\hat{\rho}$ obtained and the results given in Johnston (1963) (p. 191), we conjecture that the improvement in efficiency would be small.

Considering all the likely specification errors together, it must be concluded that, if serial correlation is present, only values of a t- or F-test statistic much greater than the critical values given above can be regarded as refuting the hypothesis under test at the given significance level; for values only slightly greater, a test is inconclusive. As long as the evidence on serial correlation suggests bias of the same order of magnitude in all the relevant standard errors, the t- and F-values may, however, still be used for comparing the explanatory power of different coefficients.

(15) As given for example in Johnston (1963) p. 194.

CHAPTER 6

ESTIMATION AND TESTING: RESULTS

6.1 Introduction

The arguments of earlier chapters have led to a final form of the equation for demand on unrestricted tariffs, previously given as (4.4.10), which may be written as

$$Y^u = \beta_1 + \sum_{a=1}^{4} Z^a [\beta_{2,a} + \beta_{3,a} \ln W^u + \beta_{4,a} \ln (L/\bar{L}) + \beta_{5,a} Z^{o1} + \beta_{6,a} Z^{o2}$$
$$+ \beta_{7,a} \log_0 \ln (\bar{E}/\bar{\pi}) + \beta_{8,a} \log \ln (P^{ef}/\bar{\pi}) + \beta_{9,a} \log \ln (\bar{P}^c/\bar{\pi})$$
$$+ \beta_{10,a} G \log \ln (\bar{P}^g/\bar{\pi}) + \beta_{11,a} G] + b \qquad (6.1.1)$$

where, to recapitulate definitions given fully in Chapter 4,

Y^u	is the quantity demanded of electricity on unrestricted tariffs per consumer per day
$a=1$	denotes space-heaters
$a=2$	water-heaters
$a=3$	cookers
$a=4$	lighting (class 41) and sundry appliances (class 42)
Z^a	is the appliance ownership level
W^u	temperature, the lower of actual temperature and an estimated threshold $\hat{\tau}^u$
L	hours of daylight (and $\bar{L}$ its mean)
Z^{o1}	ownership of off-peak space-heaters
Z^{o2}	ownership of off-peak water-heaters
$\bar{E}$	total consumption expenditure per household
$\bar{\pi}$	an all-commodities price index
P^{ef}	the final rate on the unrestricted electricity tariff
$\bar{P}^c$	the price of solid fuel
$\bar{P}^g$	the price of gas
G	the proportion of electricity consumers having a gas supply available
lag	an operator giving a weighted average of current and past values of the operand.

The total effect of a change in an exogenous variable on unrestricted demand may be divided into its effects on the demands due to ownership of each of four classes of appliances. Each of these may be further analysed into three parts: the effect on utilization given the stock of appliances, which we have called the *U effect*; the effect on the stock of appliances—the *Z effect*; and the effect on utilization associated with a change in the level or composition of the stock of appliances in the same or a different class—the *Z-U effect*.

Ex hypothesi the constant β_1 is the sum of four constants measuring the average Z-U effects of changes in the stocks of each of the four classes of appliances. A coefficient $\beta_{j,a}$ measures the sum of the U effect and Z-U effect of changes in the corresponding variable on the utilization of appliances in class a, less the average Z-U effect associated with changes in the stock of appliances in this class from all causes.

The equation was estimated in three stages.

First, some prior information was used to reduce the number of parameters to be estimated. The constraints are derived in 6.2.

Second, a final set of explanatory variables was selected. This involved (i) estimating the temperature threshold parameter τ'' introduced in 4.20 (described in 6.3); and (ii) comparing results with each of the three definitions of the cooker ownership variable Z^3 given in 4.9 and the three lag distributions defined in 4.19 (described in 6.4).

Third, using results obtained at the second stage, further constraints were introduced to improve reliability (6.5). The β coefficients in the final form of the equation incorporating all the constraints, given in 6.6, were then estimated by Generalized Least Squares. The resulting values of the coefficients—together with the descriptive and test statistics defined in Chapter 5, derived estimates of annual consumption and growth rates by Area, elasticities and utilization, and predictions for the South of Scotland—are set out and discussed in 6.7-6.12.

In 6.13 a comparison with the results of other studies is attempted.

6.2 Parameters constrained to be zero

The demand equation (6.1.1) contains 41 coefficients. It is possible, by using prior information, to determine that some of them are much closer to zero than others; and necessary to reduce the number of coefficients because of the very high correlations between some of the corresponding variables. Therefore we have constrained to be zero any coefficient which the evidence suggests is close to zero absolutely and relative to the coefficients of the same exogenous variable in other utilization functions. The adverse effects will be that some variation in demand goes unexplained and that, since a variable is not eliminated from all four utilization functions, the coefficients in the functions in which it is included take up some of the (low) explanatory power of the coefficients constrained to be zero in the other functions. But these effects will be more than compensated for by the increase in reliability.

We now consider which coefficients should be so constrained for each variable in turn.

Temperature, W. The utilization of cookers depends on the air, and water supply, temperatures both because they determine the amount of heat needed to prepare a given hot meal or drink and because the number of hot meals cooked at home falls in hot weather. Similarly some appliances in the lighting and sundry class meet a demand for heat and their utilization depends on behaviour patterns which vary with the weather. But the effect of temperature on the average utilization of all appliances in each of these classes is much smaller than on the utilization of space-heaters and water-heaters. In the sundries class, the positive coefficient for refrigerators offsets the negative one for other appliances.

We therefore set

$$\beta_{3,3}=\beta_{3,4}=0 \tag{6.2.1}$$

Hours of daylight, L. Hours of daylight is omitted from the utilization functions for space-heaters, water-heaters, and cookers:

$$\beta_{4,1}=\beta_{4,2}=\beta_{4,3}=0 \tag{6.2.2}$$

This means that we are ignoring, most notably, the effect of hours of daylight on the number of people indoors wanting space-heating, but the repercussions of this should be confined to a small effect on the temperature coefficient in U^1, given that quarterly data are used.

Whilst it is necessary, for the reasons given in 3.6, to have a single utilization function for lighting and sundry appliances, it is desirable to recognize that sundries utilization depends much less on hours of daylight than lighting utilization and probably no more than the utilization of the other three classes. We therefore relax assumption 3.3.1(i) on homogeneity of the class in one respect, making instead the following assumption:

Assumption 6.2.1

$$U^{41}=\beta_{1,41}\frac{1}{Z^{41}}+\beta_{2,41}+\ldots+\beta_{4,41}\ln(L/\bar{L})+\ldots$$

$$U^{42}=\beta_{1,42}\frac{1}{Z^{42}}+\beta_{2,42}+\ldots+\beta_{4,42}\ln(L/\bar{L})+\ldots$$

and

(i) $\beta_{4,42}=0$

(ii) $\beta_{j,41}=\beta_{j,42}$ $j=2, 3, 5, 6, \ldots, 11$ (6.2.3)

(iii) $\bar{L}=\dfrac{1'V^{-1}L}{1'V^{-1}1}$

where $\mathbf{L}$ is the vector of 624 moving-quarter values L_{qB}, and $\mathbf{V}$ and $\mathbf{1}$ are as defined in Chapter 5.

Subscripts and superscripts 41 denote lighting, 42 sundry appliances.

Writing

$$\beta_{1,4} \equiv \beta_{1,41} + \beta_{1,42}$$

$$\beta_{j,41} = \beta_{j,42} \equiv \beta_{j,4} \qquad j=2, 3, 5, 6, \ldots, 11$$

and $\qquad \beta_{4,41} \equiv \beta_{4,4}$

the effect on equation (6.1.1) is to replace the term

$$Z^4 \beta_{4,4} \ln (L/\bar{L})$$

by $\qquad Z^{41} \beta_{4,4} \ln (L/\bar{L})$

Off-peak appliance ownership, Z^{o1}, Z^{o2}. The exceptional Z-U effects associated with changes in ownership of a particular class of off-peak appliances should virtually all be effects on the utilization of the corresponding class of unrestricted appliances.

We therefore set

$$\beta_{5,2} = \beta_{5,3} = \beta_{5,4} = 0 \qquad (6.2.4)$$

and $\qquad \beta_{6,1} = \beta_{6,3} = \beta_{6,4} = 0 \qquad (6.2.5)$

The price of solid fuel, P^c. A change in the price of solid fuel must have had a much smaller U effect on the utilization of cookers than on the utilization of space-heaters and water-heaters because of the relative rarity of owning both a solid-fuel and an electrical appliance. The Z-U effects were probably also small for cookers: in most Areas ownership of solid-fuel cookers was low, and there is little evidence of a relationship between the price of solid fuel and ownership of an electric cooker across Areas.

We therefore set

$$\beta_{9,3} = 0 \qquad (6.2.6)$$

For many types of appliances within the lighting and sundry class there was no solid-fuel substitute so that Z-U effects on the mean utilization of the whole class would have been small. If there was a substitute, it was rare to own both it and the electrical appliance so that U effects would also have been small.

We therefore set

$$\beta_{9,4} = 0 \qquad (6.2.7)$$

The price of gas, P^g, *and availability of gas,* G. For the lighting and sundry class, the same conclusions about the size of Z-U and U effects apply to the price of gas as to the price of solid fuel.

We set

$$\beta_{10,4}=0 \tag{6.2.8}$$

and therefore

$$\beta_{11,4}=0 \tag{6.2.9}$$

In consequence of introducing these constraints, 26 β coefficients remain to be estimated in equation (6.1.1).

6.3 Estimation of the temperature threshold parameter, τ^u

The temperature threshold parameter τ^u can be estimated prior to the full set of β coefficients, and before selection of the final set of explanatory variables: as the temperature variable is virtually uncorrelated with all the other exogenous variables except hours of daylight, the best estimate of τ^u may be determined from an equation from which all these other variables are excluded. The equation used was

$$Y^u = b_1 + b_{2,1}Z^1 + b_{2,2}Z^2 + b_{2,3}Z^3 + b_{2,4}Z^4 + b_{4,4}Z^{41} \ln (L/\bar{L})$$
$$+ b_{3,1}Z^1 \ln W^u + b_{3,2}Z^2 \ln W^u + e \tag{6.3.1}$$

where $Z^3 = (Al)^3$, the installed load of cookers.

This was estimated by Generalized Least Squares for values of $\hat{\tau}^u$ at intervals of 0·1 °C. The results in Table 6.3.1 indicate the general pattern—all the statistics changed smoothly with $\hat{\tau}^u$.

The conclusion drawn is that the best estimate of the mean quarterly temperature above which mean quarterly unrestricted demand was virtually constant is 11·7 °C when temperature is measured as the mean at 09.00 and 21.00 hours. Hence in all further work we have used $\hat{\tau}^u = 11·7$ °C. This is the value which minimizes the seasonal pattern of the residuals as measured by

TABLE 6.3.1. *Unrestricted demand: results with different estimates of the temperature threshold*

Equation estimated: (6.3.1)

$\hat{\tau}^u$ °C	$\infty^{(1)}$	13·0	11·7	11·0
$b_{3,1}$	−4·50	−5·58	−6·18	−6·59
$b_{3,2}$	0·90	1·84	2·10	2·31
$F\{b_{3,1}^o\}$	1081	1118	1093	1055
$F\{b_{3,2}^o\}$	1·53	4·81	4·96	5·09
R^2	0·930	0·931	0·930	0·929
$t\{\beta\}$	18·11	16·32	14·91	14·21
$F\{c_5^o, \ldots, c_{15}^o\}$	19·03	16·79	16·57	16·30
$F\{c_2^o, c_3^o, c_4^o\}$	20·36	3·67	0·77	1·72

(1) These results apply for all $\hat{\tau}^u \geqslant 18.9$.

$F\{c_2^o, c_3^o, c_4^o\}$. The total sum of squared residuals is very slightly above the minimum attained at $\hat{\tau}^u = 13 \cdot 0$. The seasonal pattern has been minimized because this has the effect of giving more weight to observations of extreme values of the mean quarterly temperature than minimization of the (unweighted) sum of squared residuals does, and hence of counteracting the effect of ignoring variability within months in biasing $b_{3,1}$ and $\hat{\tau}^u$ upwards.

A mean temperature of 11·7 °C, or 53 °F, at 09.00 and 21.00 hours corresponds roughly to a noon temperature of 60 °F, a figure which has been used as threshold in some research done in the fuel industries.

The value of $F\{c_2^o, c_3^o, c_4^o\}$ for $\hat{\tau}^u = \infty$ is much higher than at $\hat{\tau}^u = 11 \cdot 7$ and none of the other statistics suggest that there is any error in specifying a threshold temperature or in using 11·7 °C: the temperature coefficients $b_{3,1}$ and $b_{3,2}$ are plausible,[1] though the fitting of a third parameter means that they are probably taking up more of the explanation of the seasonal pattern due to omitted natural variables. The evidence of the autocorrelation coefficient $\hat{\rho}$ and the Area pattern of the residuals as measured by $F\{c_5^o, ..., c_{15}^o\}$ does not refute the hypothesis that the threshold is the same in all Areas—a conclusion confirmed by examination of the d statistics for each Area.

Thus the results fulfil the expectation expressed in Chapter 2 that allowance for non-linearity of response to temperature changes would be a worthwhile improvement in the specification of the demand equation.

6.4 Selection of the cooker ownership variable, Z^3, and lag distribution

As the next part of the second stage of the analysis, equation (6.1.1) was estimated subject to the constraints (6.2.1)-(6.2.9), using in turn each of the alternative definitions of the cooker ownership variable Z^3 given in 4.9 and alternative lag distributions given in 4.19.

Our aims at this stage are, firstly, to examine the validity of the different hypotheses about cooker ownership and the lag distribution and to select the set of variables to be used in further analysis; and, secondly, to obtain estimates of the identifiable part of each utilization function for use at the next stage of the analysis in imposing constraints on the coefficients to obtain more reliable individual coefficients. Given the high correlations between the 26 explanatory variables, the individual coefficients obtained at this stage are inevitably unreliable, but our aims can be achieved by considering only summary statistics and combinations of coefficients.

We define

$$\hat{T}^a \equiv \sum_{j=2}^{11} b_{j,a} X_j \qquad (6.4.1)$$

_____

(1) Further discussed in 6.10 below.

TABLE 6.4.1. *Results with different cooker ownership variables and lag distributions*

	Z^3	A^3		$(A\bar{I})^3$			$(A\hat{I})^3$		
Lag weights	w_1	w_0	w_4	w_1	w_0	w_4	w_1	w_0	w_4
b_1	−2·04	−0·93	−1·56	−1·90	−0·77	−1·40	−1·46	−0·35	−1·03
$\bar{T}^1$	0·320*	0·157	0·179	0·229	0·120	0·122	0·026	0·152	−0·010
$\bar{T}^2$	1·13*	1·05	1·31	1·19	1·11	1·36	1·61	1·32	1·74
$\bar{T}^3$	4·83*	2·35	3·36	0·628	0·243	0·396	−0·031	−0·222	−0·212
$\bar{T}^4$	1·51*	1·52	1·52	1·60	1·55	1·58	1·84	1·63	1·80
R^2	0·967	0·970	0·967	0·967	0·970	0·967	0·967	0·970	0·967
$t\{\beta\}$	4·92	3·13	3·85	4·91	3·02	3·79	5·04	3·25	4·20

Note: $\bar{T}^a$ is measured in hours per kilowatt per day except $\bar{T}^3$ for $Z^3 = A^3$ which is in kilowatt-hours per consumer having per day.
* The figures used in the subsequent analysis.

where $b_{j,a} \equiv 0$ if $\beta_{j,a}$ is constrained to be zero, and is the regression estimate of $\beta_{j,a}$ otherwise. An estimate of utilization is then

$$\hat{U}^a \equiv \hat{T}^a + b_{1,a} \frac{1}{Z^a} \qquad (6.4.2)$$

where $b_{1,a}$ is an estimate of $\beta_{1,a}$ satisfying $\sum_{a=1}^{4} b_{1,a} = b_1$. The coefficients $b_{1,a}$ cannot be evaluated from the equation estimated. The coefficient b_1 and the values, $\bar{T}^a$, of $\hat{T}^a$ computed at the sample means of the variables X_j are given together with the other summary statistics in Table 6.4.1 for each of the nine regressions.

Three alternative definitions of cooker ownership were put forward in 4.9:

$Z^3 = (AI)^3$ the installed load of cookers per consumer,
$Z^3 = (A\tilde{I})^3$ a weighted average of the proportions owning each type of cooker,
$Z^3 = A^3$ the proportion owning an electric cooker.

The three series are highly correlated, as shown by Table 6.4.2. Consequently R^2 and $t\{\hat{\rho}\}$ are virtually the same for all three regressions, whichever set of weights is used. Some tentative conclusions can be drawn from the $\hat{T}^a$ figures. When $Z^3 = (AI)^3$—similar conclusions apply when $Z^3 = (A\tilde{I})^3$—the quarterly series for $\hat{T}^a$ show that $b_{1,a}$ must be large and positive to make $\hat{U}^a$ positive, for both $a = 1$ and $a = 3$. Hence $b_{1,2} + b_{1,4}$ must be a larger negative number to make b_1 negative. When $Z^3 = A^3$, $b_{1,1}$ must still be positive but need not be so large, and in this case it is plausible that $b_{1,3}$ is near zero. Hence $b_{1,2}$ and $b_{1,4}$ are not necessarily large. *Thus explanation of the results with $Z^3 = A^3$ depends least on the existence of large Z-U effects. For this reason cooker ownership is defined as the proportion owning an electric cooker, $Z^3 = A^3$, in further analysis.* Because of the high correlations between $(AI)^3$, $(A\tilde{I})^3$, and A^3, the adverse effects on subsequent results if A^3 is the wrong variable will be slight. The results with the other definitions of Z^3 are consistent with

TABLE 6.4.2. *Zero-order correlation coefficients: ownership variables*

	$(AI)^1$	$(AI)^2$	$(AI)^3$	$(A\tilde{I})^3$	A^3	$(AI)^{41}$	$(AI)^{42}$
$(AI)^1$	1·000	—	—	—	—	—	—
$(AI)^2$	·818	1·000	—	—	—	—	—
$(AI)^3$	·642	·855	1·000	—	—	—	—
$(A\tilde{I})^3$	·517	·778	·970	1·000	—	—	—
A^3	·452	·736	·949	·994	1·000	—	—
$(AI)^{41}$	·917	·732	·552	·446	·386	1·000	—
$(AI)^{42}$	·721	·863	·927	·840	·817	·620	1·000

TABLE 6.4.3. *Unrestricted demand: results with 26 variables,*

Variable k	b_k	$\hat{\sigma}_k$	$F\{b_k\}$	$F\{b_k^0\}$
1	$-2 \cdot 04$	$0 \cdot 442$	$21 \cdot 30$	114767
Z^1	$39 \cdot 3$	$14 \cdot 9$	$6 \cdot 97$	8516
Z^2	$-33 \cdot 7$	$22 \cdot 8$	$2 \cdot 18$	$348 \cdot 6$
Z^3	$-85 \cdot 5$	$80 \cdot 9$	$1 \cdot 12$	$0 \cdot 00$
Z^4	$1 \cdot 05$	$9 \cdot 85$	$0 \cdot 01$	$6 \cdot 67$
$Z^1 \ln W^u$	$-6 \cdot 18$	$0 \cdot 333$	$345 \cdot 4$	7968
$Z^2 \ln W^u$	$1 \cdot 87$	$0 \cdot 661$	$7 \cdot 97$	$5 \cdot 45$
$Z^{4\,1} \ln(L/L)$	$-1 \cdot 43$	$0 \cdot 218$	$43 \cdot 16$	$45 \cdot 09$
$Z^1 Z^{01}$	$-2 \cdot 76$	$0 \cdot 229$	$145 \cdot 6$	$185 \cdot 0$
$Z^2 Z^{02}$	$8 \cdot 33$	$1 \cdot 90$	$19 \cdot 14$	$13 \cdot 51$
$Z^1 \, \text{lag}_0 \, \ln (\bar{E}/\bar{\pi})$	$1 \cdot 50$	$2 \cdot 85$	$0 \cdot 27$	$64 \cdot 51$
$Z^2 \, \text{lag}_0 \, \ln (\bar{E}/\bar{\pi})$	$-4 \cdot 67$	$4 \cdot 40$	$1 \cdot 12$	$99 \cdot 11$
$Z^3 \, \text{lag}_0 \, \ln (\bar{E}/\bar{\pi})$	$18 \cdot 0$	$14 \cdot 8$	$1 \cdot 48$	$21 \cdot 78$
$Z^4 \, \text{lag}_0 \, \ln (\bar{E}/\bar{\pi})$	$0 \cdot 202$	$1 \cdot 73$	$0 \cdot 01$	$6 \cdot 79$
$Z^1 \, \text{lag}_1 \, \ln (P^{ef}/\bar{\pi})$	$0 \cdot 419$	$1 \cdot 29$	$0 \cdot 11$	$15 \cdot 22$
$Z^2 \, \text{lag}_1 \, \ln (P^{ef}/\bar{\pi})$	$-2 \cdot 12$	$1 \cdot 99$	$1 \cdot 14$	$0 \cdot 79$
$Z^3 \, \text{lag}_1 \, \ln (P^{ef}/\bar{\pi})$	$32 \cdot 0$	$8 \cdot 46$	$14 \cdot 35$	$61 \cdot 45$
$Z^4 \, \text{lag}_1 \, \ln (P^{ef}/\bar{\pi})$	$-3 \cdot 40$	$1 \cdot 10$	$9 \cdot 61$	$4 \cdot 71$
$Z^1 \, \text{lag}_1 \, \ln (\bar{P}^c/\bar{\pi})$	$-3 \cdot 28$	$1 \cdot 00$	$10 \cdot 77$	$121 \cdot 8$
$Z^2 \, \text{lag}_1 \, \ln (\bar{P}^c/\bar{\pi})$	$10 \cdot 6$	$2 \cdot 04$	$26 \cdot 92$	$32 \cdot 12$
$Z^1 \, G \, \text{lag}_1 \, \ln (\bar{P}^g/\bar{\pi})$	$-2 \cdot 06$	$1 \cdot 09$	$3 \cdot 59$	$41 \cdot 84$
$Z^2 \, G \, \text{lag}_1 \, \ln (\bar{P}^g/\bar{\pi})$	$1 \cdot 67$	$1 \cdot 92$	$0 \cdot 76$	$1 \cdot 85$
$Z^3 \, G \, \text{lag}_1 \, \ln (\bar{P}^g/\bar{\pi})$	$11 \cdot 0$	$5 \cdot 09$	$4 \cdot 68$	$0 \cdot 00$
$Z^1 \, G$	$13 \cdot 5$	$4 \cdot 68$	$8 \cdot 30$	$15 \cdot 43$
$Z^2 \, G$	$-2 \cdot 20$	$7 \cdot 53$	$0 \cdot 09$	$1 \cdot 00$
$Z^3 \, G$	$-61 \cdot 1$	$21 \cdot 3$	$8 \cdot 21$	$8 \cdot 21$

$$R^2 = 0 \cdot 967 \qquad\qquad F\{R^2\} = 703 \cdot 4$$

$Z^3 = A^3$ being correct: for, as A^3 and l^3 rose together over time, $b_{1,3}$ would be large and positive if $(Al)^3$ were used when A^3 was correct.

Similarly, because of high correlations between the lagged variables derived from any given price series, only a tentative conclusion can be reached on which is the correct lag distribution. Three alternatives were defined in 4.19:

weights w_0 implying a linear adjustment path

$\qquad\quad$ w_1 from a lognormal distribution with mode $1 \cdot 5$ implying a maximum rate of adjustment in the quarter following that in which the price changed,

$\qquad\quad$ w_4 from a lognormal distribution with mode $4 \cdot 5$ implying a maximum rate of adjustment one year after a price change.

Weights w_1 from a lognormal distribution with mode $1 \cdot 5$ are used subsequently because the lagged variables appear to explain U effects best with

proportion owning as the cooker ownership variable, and lag weights w_1

Dummy variable k	c_k	$F\{c_k\}$	$F\{c_k^o\}$	d_B
1	0·478	0·49	0·01	
Moving-quarter:				
II	−0·0638	0·02	0·40	
III	−0·201	0·16	0·18	
IV	−0·767	2·37	2·37	
Area:				
South Western	—	—	—	1·99
London	−0·160	0·03	0·00	1·92
South Eastern	−2·10	5·94	10·63	1·30
Southern	1·76	4·13	9·35	1·74
Eastern	0·0502	0·00	0·21	1·93
East Midlands	−0·447	0·27	0·12	1·58
Midlands	1·21	1·96	6·37	1·38
South Wales	−1·88	4·72	6·50	1·64
Merseyside and North Wales	0·563	0·43	1·80	1·34
Yorkshire	0·545	0·40	2·89	0·87
North Eastern	0·518	0·36	5·52	1·07
North Western	−2·48	8·23	8·23	1·78

$$F\{c_2^o, c_3^o, c_4^o\} = 0·98$$
$$F\{c_5^o, \ldots, c_{15}^o\} = 4·69$$

Autocorrelation coefficients: $\hat{\rho} = 0·195$ $\quad t\{\hat{\rho}\} = 4·92$

$\hat{\rho}_a = 0·562$ $\quad t\{\hat{\rho}_a\} = 7·57$

Heteroscedasticity coefficient: $\hat{\psi} = 0·201$ $\quad t\{\hat{\psi}\} = 1·82$

these weights: that $|b_1|$ is larger with weights w_1 than with w_0 or w_4, and the serial correlation slightly stronger, suggests that the lagged variables explain less of the Z-U effects with these weights—as one would expect since the other distributions give more weight to distant values; but R^2 is virtually the same.

Table 6.4.3 gives the complete results with $Z^3 = A^3$ and weights w_1 for comparison with later results.

6.5 Constraints on elasticities

In order to obtain more reliable coefficients than given by the 26-variable regression, it is necessary to reduce further the number of terms containing the same exogenous variables. A convenient starting-point in investigating what relationship may most reasonably be assumed to hold between the coefficients in different utilization functions is an expression for the elasticity of utilization with respect to a given variable.

Let us consider first total consumption expenditure. Suppose an increment in $\text{lag}_0 (\bar{E}/\bar{\pi})_q$ is brought about by an equiproportionate increment in each component $(\bar{E}/\bar{\pi})_{q+t}$, $t = -15, \ldots, 0$, $\bar{\pi}_{q+t}$ remaining constant all t. Then the elasticity of the utilization of appliances in class a with respect to $\text{lag}_0 \, (\bar{E}/\bar{\pi})$ is

$$\frac{\text{lag}_0 \, (\bar{E}/\bar{\pi})}{U^a} \frac{\partial U^a}{\partial \, \text{lag}_0(\bar{E}/\bar{\pi})} = -\beta_{1a}\left(\frac{1}{Z^a U^a}\right)\frac{\text{lag}_0 \, (\bar{E}/\bar{\pi})}{Z^a}\frac{\partial Z^a}{\partial \text{lag}_0(\bar{E}/\bar{\pi})} + \frac{\beta_{7a}}{U^a}$$

(6.5.1)

assuming appliance ownership Z^a is continuously differentiable.

This measures the total effect over sixteen quarters of a permanent change in $\bar{E}/\bar{\pi}$. The first term on the right-hand side is the elasticity due to the average Z-U effect of an increment in Z^a from any cause, times the elasticity of Z^a with respect to $\text{lag}_0(\bar{E}/\bar{\pi})$; the second term is the elasticity due to the U effect plus any Z-U effect different from the average.

If the expenditure coefficients β_{7a} were constrained to equality for all classes a, it would imply that any change in real consumption expenditure $\bar{E}/\bar{\pi}$ brought about the same absolute increase in the utilization of all classes of appliances, or in other words, that the elasticity was inversely related to utilization, over classes. This form of constraint might be realistic if a given class's having a low value of U^a relative to those for other classes means that utilization of that class was relatively far below some saturation level. But in view of the differing consumption characteristics of the classes, and the difference in the units of measurement of U^3 when $Z^3 = A^3$, it seems likely to be nearer the truth to say that the elasticities are the same at the sample mean, that is

$$\frac{\beta_{7a}}{U^a} = \bar{\beta}_7^* \qquad \text{all } a \tag{6.5.2}$$

U^a is defined as

$$\bar{U}^a \equiv \sum_{j=1}^{11} \beta_{j,a}\bar{X}_j$$

where

$$\bar{X}_1 = 1/\bar{Z}^a$$

and $\bar{X}_j(j = 2, 3, \ldots, 11)$ and $\bar{Z}^a$ are means of the 624 sample observations.

To introduce a constraint of this form it is necessary to replace the unknown $\bar{U}^a$ by a quantity known prior to estimation of the constrained equation. The relationship assumed is

$$\frac{\beta_{7,a}}{\bar{T}^a} = \bar{\beta}_7 \qquad \text{all } a \tag{6.5.3}$$

where the $\bar{T}^a$ are the values of the identifiable part of the utilization function obtained from the 26-variable regression with $Z^3 = A^3$ and weights w_1.[2]

If (6.5.2) is the true relationship, no systematic error will arise from

(2) The starred values in Table 6.4.1.

using (6.5.3) as long as the average Z-U effect is the same for all classes at the mean $\bar{U}^a$ when expressed as an elasticity, that is, as long as

$$\frac{\bar{Z}^a}{\bar{U}^a}\frac{\partial U^a}{\partial Z^a}=\frac{-\beta_{1,a}}{\bar{Z}^a\bar{U}^a}$$

is the same for all a.

From a similar argument, the following relationships are postulated amongst the price parameters:

$$\frac{\beta_{8,a}}{\bar{T}^a}=\bar{\beta}_8 \qquad a=1, 2, 3, 4 \qquad (6.5.4)$$

$$\frac{\beta_{9,a}}{\bar{T}^a}=\bar{\beta}_9 \qquad a=1, 2 \qquad (6.5.5)$$

$$\frac{\beta_{10,a}}{\bar{T}^a}=\bar{\beta}_{10} \qquad a=1, 2, 3 \qquad (6.5.6)$$

(6.5.6) implies the following relationship between the coefficients of gas availability:

$$\frac{\beta_{11,a}}{\bar{T}^a}=\bar{\beta}_{11} \qquad a=1, 2, 3 \qquad (6.5.7)$$

To sum up, our hypothesis is that the elasticity of utilization with respect to real consumption expenditure, and the prices of electricity, solid fuel and gas is the same for each class of appliance when calculated at the sample mean just defined.

6.6 The constrained demand equation

On incorporating in equation (6.1.1) the constraints of 6.2 and 6.5, and using the results of 6.3 and 6.4, the demand equation becomes

$$\begin{aligned}
Y^u={}&\beta_1+\beta_{2,1}Z^1+\beta_{2,2}Z^2+\beta_{2,3}Z^3+\beta_{2,4}Z^4\\
&+\beta_{3,1}Z^1\ln W^u+\beta_{3,2}Z^2\ln W^u+\beta_{4,4}Z^{41}\ln(L/\bar{L})\\
&+\beta_{5,1}Z^1Z^{o1}+\beta_{6,2}Z^2Z^{o2}\\
&+\bar{\beta}_7K^4\,\mathrm{lag}_0\,\ln(\bar{E}/\bar{\pi})+\bar{\beta}_8K^4\,\mathrm{lag}_1\,\ln(P^{ef}/\bar{\pi})\\
&+\bar{\beta}_9K^2\,\mathrm{lag}_1\,\ln(\bar{P}^c/\bar{\pi})+\bar{\beta}_{10}K^3G\,\mathrm{lag}_1\,\ln(\bar{P}^g/\bar{\pi})+\bar{\beta}_{11}K^3G\\
&+\varepsilon
\end{aligned} \qquad (6.6.1)$$

where $Z^3=A^3$, the proportion owning cookers,

$\qquad W_m^u=\min(W_m, 11\cdot7^\circ C)$

and $\qquad K^k=\sum_{a=1}^{k}Z^a\bar{T}^a \qquad k=2, 3, 4 \qquad (6.6.2)$

The results of estimating this by Generalized Least Squares are given in Table 6.6.1.

TABLE 6.6.1. *Unrestricted demand: results*

Variable k	b_k	$\hat{\sigma}_k$	$F\{b_k\}$	$F\{b_k^0\}$
1	0·616	0·288	4·57	96788
Z^1	31·4	2·26	192·7	7182
Z^2	−32·0	4·67	47·08	294·0
Z^3	−17·7	2·08	73·08	0·00
Z^4	−3·00	0·778	14·91	5·62
$Z^1 \ln W^u$	−6·06	0·360	283·3	6720
$Z^2 \ln W^u$	1·75	0·717	5·92	4·60
$Z^{41} \ln (L/L)$	−1·51	0·235	41·50	38·02
$Z^1 Z^{01}$	−3·27	0·208	248·6	156·0
$Z^2 Z^{02}$	6·61	1·89	12·18	11·39
$K^4 \lag_0 \ln (\bar{E}/\bar{\pi})$	0·481	0·0867	30·77	18·49
$K^4 \lag_1 \ln (P^{ef}/\bar{\pi})$	−0·203	0·0467	18·86	2·35
$K^2 \lag_1 \ln (\bar{P}^c/\bar{\pi})$	3·66	0·248	216·8	282·6
$K^3 G \lag_1 \ln (\bar{P}^g/\bar{\pi})$	−0·171	0·107	2·54	1·95
$K^3 G$	1·24	0·420	8·77	8·77

$$R^2 = 0·960 \qquad\qquad F\{R^2\} = 1052$$

TABLE 6.7.1. *Consumption on unrestricted tariffs by Area*

Actual and predicted figures (predicted figures are in brackets)

	Unrestricted units billed per domestic consumer			
	1955/56	Growth rates		1967/68
Electricity Board		'55/56 to '61/62	'64/65 to '67/68	
	kWh	% p.a.	% p.a.	kWh
London	1617 (1565)	8·1 (6·9)	0·8 (1·9)	3098 (3115)
South Eastern	1901 (1829)	7·1 (7·5)	0·9 (0·8)	3469 (3880)
Southern	1711 (1673)	9·8 (9·2)	0·4 (0·1)	3553 (3631)
South Western	1586 (1526)	8·4 (8·1)	1·9 (−1·1)	3336 (3086)
Eastern	1818 (1869)	7·9 (7·1)	0·9 (−1·1)	3468 (3481)
East Midlands	1458 (1337)	7·3 (8·1)	1·8 (2·3)	2873 (3068)
Midlands	1640 (1561)	8·5 (8·8)	0·8 (1·4)	3423 (3440)
South Wales	1209 (1404)	8·8 (6·4)	3·2 (0·8)	2662 (2597)
Merseyside and North Wales	1483 (1372)	8·8 (10·1)	2·2 (3·3)	3322 (3474)
Yorkshire	1297 (1403)	11·1 (7·5)	0·0 (3·7)	2949 (3104)
North Eastern	1104 (1194)	9·7 (6·5)	2·6 (4·0)	2497 (2553)
North Western	1445 (1549)	8·8 (7·7)	1·3 (1·4)	3040 (3169)
England and Wales	1550 (1549)	8·6 (7·8)	1·2 (1·3)	3175 (3265)

with the constrained equation (6.6.1)

Dummy variable k	c_k	$F\{c_k\}$	$F\{c_k^o\}$	d_B
1	2·10	8·21	0·01	
Moving-quarter:				
II	0·0115	0·00	0·98	
III	−0·322	0·36	0·10	
IV	−0·938	3·09	3·09	
Area:				
South Western	—	—	—	1·57
London	1·20	1·70	22·47	1·85
South Eastern	−4·65	25·28	17·62	0·79
Southern	−1·28	1·92	0·62	1·63
Eastern	−2·89	9·76	2·98	1·73
East Midlands	−1·94	4·38	0·16	1·39
Midlands	−0·129	0·02	6·37	1·27
South Wales	−3·73	16·26	9·26	1·58
Merseyside and North Wales	−1·65	3·19	0·03	1·58
Yorkshire	−0·77	0·69	1·78	0·74
North Eastern	−0·98	1·12	2·21	0·90
North Western	−4·35	22·08	22·08	1·37

$$F\{c_1^o, c_3^o, c_4^o\} = 1·39 \qquad F\{c_5^o, ..., c_{15}^o\} = 7·78$$

Autocorrelation coefficients: $\hat{\rho} = 0·295$ $t\{\hat{\rho}\} = 7·55$

$\hat{\rho}_a = 0·612$ $t\{\hat{\rho}_a\} = 8·45$

Heteroscedasticity coefficient: $\hat{\psi} = 0·320$ $t\{\hat{\psi}\} = 2·78$

TABLE 6.7.2. *The seasonal pattern of unrestricted consumption in the South West*

	Unrestricted units billed per domestic consumer: ratio of South West figure to England and Wales mean: geometric means for the period 1955II-1968I	
Quarter	Actual	Predicted by equation (6.6.1)
I	0·985	0·950
II	1·021	0·987
III	1·068	1·060
IV	1·045	1·022
Year	1·029	1·004

6.7 Goodness of fit

The results show that there were large variations in demand which were not due to variations in appliance ownership, utilization remaining constant, and that these Z-U and U effects are well explained by equation (6.6.1).

When defining, in Chapter 1, the question which we were setting out to answer, we gave in Table 1.3.1 annual consumption levels and growth rates by Area. The extent to which equation (6.6.1) explains the variations in the unrestricted part of demand may be seen from Table 6.7.1: it explains fairly well the absolute level of unrestricted consumption per consumer in each Area and the downturn in the rate of growth in the latter part of the period, but explanation of the absolute levels of the growth rates in each part of the period is poor for some Areas—as could be expected given our conclusions on the reliability of the appliance ownership series in 4.11.

The results with (6.6.1) confirm those reported in 6.3: with just four parameters it is possible to explain virtually all the seasonal pattern when averaged over Areas. The equation also explains some of the variation between Areas in the seasonal pattern. For example it correctly gives that the variation in unrestricted demand between winter and summer in the South West, as described by the figures in Table 1.3.2, was relatively small— see Table 6.7.2.

6.8 The pattern of the residuals

But the residuals do display a seasonal pattern in some Areas. The reasons for this are probably (1) errors in the temperature variable because the recording stations used were unrepresentative[3]—errors to which the results are particularly sensitive because of the high temperature coefficients; (2) errors in the monthly weights M_q^m because of variations in meter-reading behaviour between Boards;[4] (3) heterogeneity of the space-heating and water-heating classes, especially differences in temperature sensitivity between main and supplementary water-heaters;[5] and (4) omission of temperature from the cooker utilization function.[6]

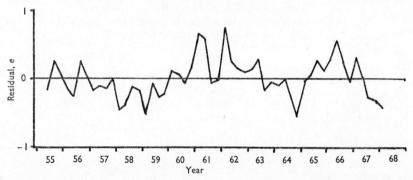

FIG. 6.8.1. The residuals, e, from estimating equation (6.6.1): England and Wales average.

(3) See 4.20.
(4) See 4.6.
(5) See the discussion of assumption 3.3.1 in 3.6, and 4.20.
(6) See 6.2 above.

There is also some residual Area pattern: $F\{c_5^o, \ldots, c_{15}^o\} = 7\cdot78$. Errors in the temperature variable,[7] in deriving Area figures for total consumption expenditure from Standard Region ones,[8] and in using the same price index π for all Areas,[9] and sampling errors in the surveys of appliance ownership[10] probably account for this, though the omission of arguments of the appliance ownership functions Z^a, such as taste variables and longer lagged variables, which cause Z-U effects may be a contributory factor.

There is strong evidence of positive serial correlation not due to a residual seasonal pattern, and varying between Areas. Figure 6.8.1 shows the pattern in the residuals e for England and Wales. The most obvious explanation of the pattern with peaks in 1955, 1961, and 1966 is errors in the appliance ownership series due to interpolating between, and extrapolating from, the figures given by the Electricity Council surveys in these years.

Whilst there is some evidence of heteroscedasticity, the hypothesis used does appear to be nearer the truth than the alternative that the standard deviation of an error is proportional to the expected value of the dependent variable, or some measure of appliance ownership, as implied by the rejected expression (3.10.1).

6.9 Estimated elasticities of unrestricted demand

It is useful to derive elasticities of unrestricted demand from the elasticities of utilization before discussing the coefficients.

Let us consider first a change in the weighted average of current and past values of real total consumption expenditure $\text{lag}_0(\bar{E}/\bar{\pi})$ brought about by an equiproportionate increment in each component as defined in 6.5.

Let $\bar{Z}^a$ and $\bar{X}_j$ be the sample means of ownership of appliances in class a and the jth determining variable respectively, and in particular $\bar{X}_7$ be the sample mean of $\text{lag}_0 \ln(\bar{E}/\bar{\pi})$.

Define
$$\overline{\text{lag}_0(\bar{E}/\bar{\pi})} \equiv \exp(\bar{X}_7)$$

and a mean of the quantity demanded as

$$\tilde{Y}^u \equiv b_1 + \sum_a \bar{Z}^a \bar{T}^a, \qquad (6.9.1)$$

calculated from the 26-variable regression results in Tables 6.4.1 and 6.4.3.

Then the elasticity of unrestricted demand with respect to consumption expenditure $\bar{E}$, at the mean just defined, is

$$\frac{\overline{\text{lag}_0(\bar{E}/\bar{\pi})}}{\tilde{Y}^u} \frac{\partial Y^u}{\partial \text{lag}_0(\bar{E}/\bar{\pi})} = \bar{\beta}_7 \frac{\sum\limits_{a=1}^{4} \bar{Z}^a \bar{T}^a}{\tilde{Y}^u} + \tilde{\xi}_7 \qquad (6.9.2)$$

(7) See 4.20.
(8) See 4.17.
(9) See 4.18.
(10) See 4.11.

where ξ_7 represents terms in the partial derivatives of appliance ownership,

$$\frac{\partial Z^a}{\partial \, \mathrm{lag}_0(\bar{E}/\bar{\pi})} \; (a=1, \ldots, 4) \text{ and } \frac{\partial Z^{oa}}{\partial \, \mathrm{lag}_0(\bar{E}/\bar{\pi})} \; (a=1, 2)$$

and partial derivatives are evaluated at the mean.

The first term on the right-hand side, which we shall call the *β-elasticity*, is the elasticity due to the U effect plus the difference, for each class, between the Z-U effect of a change in appliance ownership Z^a caused by an increment in $\mathrm{lag}_0(\bar{E}/\bar{\pi})$ and the average Z-U effect of all changes in Z^a of this size whatever the cause. ξ_7 is the elasticity due to the average Z-U effects, Z effects, and Z-U effects associated with changes in off-peak appliance ownership.

Similar expressions may be derived for the other variables; those for the ownership of off-peak space-heaters and water-heaters, Z^{o1} and Z^{o2}, measure the effect of changes in these variables when no other arguments of unrestricted demand change.

The estimated β-elasticities, at the mean defined above, are given in Table 6.9.1.

TABLE 6.9.1.   *Unrestricted demand: β-elasticities estimated
from equation (6.6.1)*[1]

Variable	β-elasticity	Variable	β-elasticity
$\bar{E}$	0·45	W^u	−1·20
P^{ef}	−0·19	L	−0·12
$\bar{P}^c$	0·73	Z^{o1}	−0·042
$\bar{P}^g$	−0·047	Z^{o2}	0·0038
G	0·20		

(1) Calculated at the sample means defined in 6.9. Given the rapid rise in off-peak appliance ownership, the specification of the equation implies a considerable change in the elasticities with respect to Z^{o1} and Z^{o2} over the period. For example in England and Wales in 1955/56, both were zero; in 1968I that with respect to Z^{o1} was −0·25 and that with respect to Z^{o2} +0·025. The figures quoted for other variables are much more representative averages.

6.10 The interpretation of the coefficients

When interpreting the coefficients, it is important to remember that they, and the elasticities given in Table 6.9.1, do not measure the total effect on the unrestricted demand for electricity. For example the total effect, and the

U effect and Z effect on a particular class of appliances, of an increase in total consumption expenditure could all be positive—if electricity, these appliances and their utilization were all superior; yet the sum, which *ex hypothesi* the expenditure coefficient measures, of the U effect and the deviation of the Z-U effect from the average could be negative. This sort of possibility has to be borne in mind when interpreting all the coefficients.

The total expenditure, price, and gas availability variables together explain a significant part of the U and Z-U effects: the group F-value is 62·83. The coefficients of total expenditure, the price of electricity and the price of solid fuel are also individually significant and of the same sign as the expected U effect. At least for the price of solid fuel, the elasticity in Table 6.9.1 is high to be entirely due to a U effect.

The total expenditure coefficient, b_7, appears plausible but probably includes the effects of changes in many other, omitted variables,[11] and could be biased towards zero by errors in the price index π.[12] It seems more likely that b_7 has been raised than lowered by the Z-U effects of changes in real consumption expenditure $\bar{E}/\pi$ and correlated omitted variables. These may therefore have been small and negative, or positive, since the average Z-U effect associated with an increase in the stock of appliances in all classes is negative ($b_1 > 0$).

Discussion of the interpretation of the electricity price coefficient is postponed until Chapter 8[13] as further light is thrown on this by the results with the tariff function model.

The high value of the solid-fuel price coefficient, b_9, is probably explained by three factors: firstly it is probably upward biased because the range of the price of solid fuel is underestimated;[14] secondly, as for total expenditure, the Z-U effect was greater than the negative average one: in Areas with high solid fuel prices, the higher ownership of electrical appliances was associated with only slightly lower, or possibly higher, mean utilization. This could be partly due to the solid fuel price taking up the explanation of a positive Z-U effect which would be correctly explained by expenditure, *E*, if a weighted mean were used for the latter.[15] Thirdly it has been suggested that there was a shift of tastes away from solid fuels over the period, and that the Clean Air Act led to a significant reduction in the use of solid-fuel appliances.[16] If this was so and the effect was to lead to the acquisition of electrical appliances by households which used them relatively intensively

(11) See especially the discussion of assumption 3.5.2 in 3.6.
(12) See 4.18.
(13) See 8.10.
(14) See 4.16.
(15) See 4.17.
(16) See Shell International Petroleum Co. (1969).

—most plausibly because they were used as the main form of heating—the rise in mean utilization would have been correlated with the rise in the real price of solid fuel, $\bar{P}^c/\bar{\pi}$.

The gas price coefficient, $\bar{b}_{10}$, has the opposite sign to the expected U effect, but is small and insignificant. For any one class of appliances, it could be that the U effect is zero, as implied by assumption 3.8.2, and the Z-U effect negative and only slightly larger absolutely than the average; or that there is a positive U effect and a negative Z-U effect which is larger than the average by an amount slightly greater absolutely than the U effect. Given the low value of $\bar{b}_{10}$, $\bar{b}_{11}$, the coefficient of the term in gas availability alone, is high enough to indicate the existence of large negative Z-U effects arising from variations in gas and electrical appliance ownership between Areas because of differences in gas availability.[17] The sign implies that a higher proportion of low utilization households use gas where it is available than high utilization ones. If the response to differences in price is also greater amongst low utilization households, this would explain why $\bar{b}_{10}$ is negative.

The positive sign of the temperature coefficient for water-heating, $b_{3,2}$, shows that observations in the range of temperatures in which the bringing into use of supplementary heaters produces a rise in mean utilization dominate the determination of the coefficient, and therefore implies that the variations in U^2 corresponding to variations in temperature between winter quarters in successive years are wrongly predicted.[18] The expectation expressed in 2.12 that hours of daylight would add significantly to the proportion of variation in demand explained is fulfilled.

That the coefficient of off-peak space-heating ownership, $b_{5,1}$, is large, negative, and significant supports the hypothesis[19] that the slowing down in the rate of growth of unrestricted demand from 1964/65 was partly explained by the occurrence of large negative Z-U effects on the utilization of unrestricted space-heaters from the growth of off-peak space-heating ownership—because the growth was mainly amongst households which would otherwise have had a high demand for unrestricted space-heating. This inference is drawn because $b_{1,1}$, the component of the constant measuring the average Z-U effect on space-heating, is probably, although positive,[20] smaller absolutely than $b_{5,1}$, and because any given change in off-peak space-heating ownership, Z^{o1}, would be associated with a smaller change in unrestricted space-heating ownership: some of an increase in off-peak ownership would be due to substitution for other-fuel appliances; unrestricted heaters may have been acquired by some households to complement their

(17) See 3.8.
(18) See 4.20.
(19) See 3.7.
(20) See 6.11 below.

main off-peak heaters; and the loading of off-peak installations was relatively high.

Although the coefficient of off-peak water-heating ownership, $b_{6,2}$, is absolutely higher than $b_{5,1}$, the growth of off-peak water-heating ownership was much slower than that of off-peak space-heating ownership. The coefficient therefore indicates that the mean utilization of unrestricted water-heaters was little affected by the growth of off-peak ownership. What Z-U effect there was was probably positive. The likely explanation of this sign is that more consumers with unrestricted supplementary heaters and other-fuel main heaters were changing to off-peak than consumers whose unrestricted heater was their main or only heater.

6.11 Estimates of utilization

Some tentative conclusions can be drawn from the results as to the contribution to the changes in unrestricted demand made by changes in the utilization of each class of appliances.

An estimate of the utilization of appliances in class a may be expressed with reference to the demand equation (6.1.1) as

$$\hat{U}^a = \hat{T}^a + b_{1,a}\frac{1}{Z^a} \tag{6.11.1}$$

where $\hat{T}^a$ is an estimate of the function of the explanatory variables in square brackets in (6.1.1), and $b_{1,a}$ is an estimate of the average Z-U effect of changes in the ownership of appliances in class a, Z^a, the estimate being defined such that

$$\sum_{a=1}^{4} b_{1,a} = b_1, \text{ the estimator of the constant } \beta_1.\text{[21]}$$

Estimates $\hat{T}^a$ have been computed for each class of appliances from the estimated coefficients of the constrained demand equation (6.6.1)—using the constrained values of parameters where appropriate. The four series for England and Wales are shown in Figure 6.11.1.

Utilization can only be roughly estimated because, firstly, there are no regression estimates of the individual coefficients $b_{1,a}$, only of their sum. Some clues to their values can be deduced from the requirements that they sum to the estimated value b_1, and that the estimates of utilization must be positive. As the ownership of all classes of appliances grew in England and Wales, $| \hat{U}^a - \hat{T}^a |$ declined over time. Secondly the estimates of the identifiable part, $\hat{T}^a$, are very sensitive to errors in the specification of the equation, particularly errors in the parameter constraints.

Utilization of space-heaters, U^1. It is probable that the seasonal variation

(21) See 3.7 and 6.4 above for a fuller discussion.

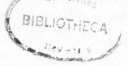

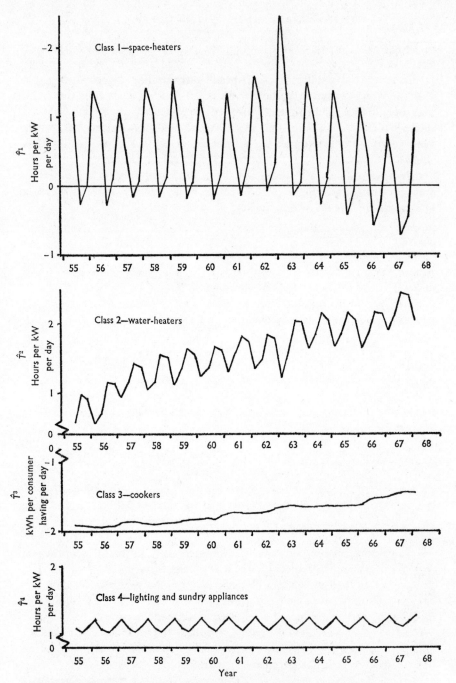

FIGURE 6.11.1. *The estimates of the identifiable parts of the unrestricted utilization functions for England and Wales from the results with equation (6.6.1)*

in $\hat{T}^1$ is overestimated because temperature was omitted from the utilization function for cookers. It is nevertheless virtually certain that $b_{1,1}$ is positive. Hence the trend in the utilization of space-heaters was probably nearly constant, or falling slowly, from 1955 to 1962, and falling more rapidly thereafter.

Utilization of water-heaters, U^2. The configuration of the $\hat{T}^2$ series and the 26-variable regression results are consistent with $b_{1,2}$ being close to zero. We therefore conclude that there was probably a fairly strong upward trend in the utilization of water-heaters, as in $\hat{T}^2$.

Utilization of lighting and sundry appliances, U^4. Similarly, there was probably a very slight upward trend in the utilization of lighting and sundry appliances as in $\hat{T}^4$.

Utilization of cookers, U^3. A comparison of the coefficients of total expenditure, prices, and gas availability from the 26-variable regression with the coefficients implied by the estimate of the constrained equation suggests that Z-U effects on the utilization of cookers were relatively large—and especially that an increase in the price of electricity had an exceptionally large positive Z-U effect; that, as a result, the imposition of the constraints leads to coefficients in the utilization function for cookers which are too low on average; and so that both the level and rate of growth of $\hat{T}^3$ are too low. A value of $b_{1,3}$ close to zero was indicated by the 26-variable regression. We therefore conclude that there was probably an upward trend in the utilization of cookers somewhat stronger than that in the $\hat{T}^3$ series as drawn.

6.12 Prediction

Unrestricted demand, Y^u, in the South of Scotland Area has been predicted for moving-quarters from 1961II to 1968I, the period for which the actual figures are available; the results are given in Table 6.12.1.

Comparison of the actual figures for the South of Scotland in Table 6.12.2 with those for Areas in England and Wales in Table 6.7.1 shows that both the level and rate of growth of annual unrestricted consumption per consumer were much higher in the South of Scotland than in any other Area. The estimated equation fails to predict this.

The explanation of the prediction errors may lie in the existence of errors in the variables—some of the series are less reliable for the South of Scotland than for other Areas—but they must be taken as strongly suggesting specification errors not indicated by the other tests.

Misspecification of the functional form could give rise to large prediction errors: as indicated in 5.9, this predictive test is a severe one because many of the exogenous variables took extreme values in the South of Scotland. In particular, the coefficient estimates imply low values of the identifiable part of the utilization function for space-heaters, and high values of the identifiable part of the function for water-heaters, in the South of Scotland because

TABLE 6.12.1. *Predictions of unrestricted demand in the South of Scotland from equation (6.6.1)*

Predicted figures are given in brackets after the actual figures, with the standard errors underneath

Moving-quarter	Kilowatt-hours per consumer per day			
	II	III	IV	I
Year				
1961/62	6·67 (6·44)	5·23 (5·26)	6·83 (6·35)	10·19 (9·35)
	0·44	0·44	0·44	0·45
1962/63	8·79 (8·21)	6·24 (5·68)	7·96 (6·54)	11·58 (10·73)
	0·45	0·45	0·44	0·45
1963/64	9·76 (9·00)	6·98 (6·07)	8·43 (7·12)	11·42 (10·42)
	0·45	0·45	0·46	0·47
1964/65	9·60 (9·12)	7·09 (6·35)	8·79 (7·69)	12·37 (11·47)
	0·45	0·48	0·48	0·49
1965/66	10·68 (9·74)	7·62 (6·42)	9·46 (7·77)	13·11 (11·51)
	0·50	0·51	0·52	0·54
1966/67	10·50 (9·20)	8·04 (6·24)	9·42 (7·96)	13·33 (11·44)
	0·56	0·57	0·57	0·59
1967/68	11·29 (9·26)	8·19 (6·48)	9·16 (7·72)	13·88 (11·72)
	0·63	0·64	0·66	0·67

TABLE 6.12.2. *Annual consumption on unrestricted tariffs in the South of Scotland*

Predicted figures are given in brackets after the actual figures

Unrestricted units billed per domestic consumer	Growth rate '64/65 to 67/68 % p.a.	1967/68 kWh
	3·7 (0·3)	3875 (3207)

of the high level of off-peak appliance ownership. Possibly the ownership levels of off-peak space-heaters and water-heaters should be included non-linearly such that constant rates of increase in them imply declining rates of change in the utilization of unrestricted space-heaters and water-heaters. Alternatively it could be that there is a stable relationship between the weighted and unweighted means of the exogenous variables in Areas of England and Wales[22] but a different relationship in the South of Scotland, either simply because the level of off-peak ownership is higher, or possibly because the main causes of the growth of off-peak ownership were different —a contributory factor may for example have been the high proportion of

(22) See the discussion of assumption 3.4.2 in 3.6.

council housing in the South of Scotland. Further evidence that the prediction errors are in part due to a higher proportion of low utilization households being owners of off-peak appliances in the South of Scotland is the overprediction by our model of off-peak demand in the South of Scotland.[23]

Another source of prediction errors could be the omission of variables. Notably, households were larger in the South of Scotland than in any other Area.

6.13 Comparison with the results of previous studies

Our β-elasticities differ from elasticities of unrestricted demand by the terms representing the Z and average Z-U effects in expressions such as (6.9.2). The elasticities of demand could differ from those obtained in other studies principally because of differences in the specification of the equation, especially the inclusion of lagged variables, the use of pooled quarterly time-series and Area data, the time-period and area considered, the definitions of the exogenous variables, and the use of unrestricted, instead of total, consumption.

To compare our results with those of other studies, and attempt to isolate some of these causes of differences, subject to the limitations imposed on the choice of alternative specifications by the data available, we estimated two more equations.

To obtain elasticities of demand most directly comparable with the estimated β-elasticities,

$$Y^u = \alpha_1 + \alpha_3 \ln W^u + \alpha_4 \ln (L/\bar{L}) + \alpha_7 \, \text{lag}_0 \, \ln (\bar{E}/\bar{\pi}) + \alpha_8 \, \text{lag}_1 \, \ln (P^{ef}/\bar{\pi})$$
$$+ \alpha_9 \, \text{lag}_1 \, \ln (\bar{P}^c/\bar{\pi}) + \alpha_{10} \, G \, \text{lag}_1 \, \ln (\bar{P}^g/\bar{\pi}) + \alpha_{11} G + \varepsilon \qquad (6.13.1)$$

was estimated by Generalized Least Squares from the pooled data, still using the quarterly time-series.

To obtain results more comparable with those in Stone (1954) and Wigley (1968),

$$\ln (Q/C) = \alpha_1' + \alpha_3' \ln W + \alpha_7' \ln (\bar{E}/\bar{\pi}) + \alpha_8' \ln (P^{ef}/\bar{\pi}) + \alpha_9' \ln (\bar{P}^c/\bar{\pi})$$
$$+ \alpha_{10}' \ln (\bar{P}^g/\bar{\pi}) + v \qquad (6.13.2)$$

was estimated from the pooled data, but after converting the time-series to annual ones. Ordinary Least Squares was used as, although the covariances between successive annual errors are positive because of the use of units billed data, they are very much smaller in relation to the variances than in the quarterly analysis. This equation was also estimated omitting the last two years, 1966/67 and 1967/68, from each of the Area time-series to remove as much as possible of the effect of a difference in the period covered from a comparison with Wigley's results.

(23) See 7.24.

The resulting elasticities of demand and summary statistics are presented in Table 6.13.1.

TABLE 6.13.1. *Comparative analysis: results with equations (6.13.1) and (6.13.2)*

Equation	(6.13.1)		(6.13.2)	
Elasticities computed at $Y^u =$ [1]	$\tilde{Y}^u$	$\bar{Y}^u$	—	—
Last year of annual time-series	—	—	1967/68	1965/66
Elasticity of demand[2] with respect to				
W [3]	−1·24	−1·71	−2·07	−1·70
L	−0·089	−0·12	—	—
$\bar{E}$	+1·05	+1·45	+1·28	+1·49
P^{ef}	−0·22	−0·30	−0·16	−0·22
$\bar{P}^c$	+1·13	+1·56	+1·80	+1·92
$\bar{P}^g$	−0·38	−0·52	−0·62	−0·75
G	−0·17	−0·23	—	—
R^2	0·887		0·910	0·922
$F\{R^2\}$	687·9		303·0	297·7
$\hat{\rho}_a$	0·76		0·68	0·59
$t\{\hat{\rho}_a\}$	10·28		9·99	7·38

(1) $\tilde{Y}^u$ is defined in 6.9. $\bar{Y}^u$ is the mean of the 624 sample ob-
    servations. The sample means of the explanatory variables
    were used.
(2) Elasticity of unrestricted demand only from equation (6.13.1).
(3) W^u for equation (6.13.1).

In all our results, the elasticity with respect to the price of gas ($\bar{P}^g$) is implausibly negative. This suggests that changes in some omitted variable or variables, such as tastes or appliance prices, caused the demands for gas and electricity to rise together over time—and to be high in the same Areas—so that a downward bias results from using the *average* marginal price of gas. This same phenomenon would tend to bias upward the coefficient of the price of electricity because the final rate, P^{ef}, has been used instead of an average. The direction of the net bias of this coefficient is, however, uncertain as the high correlation between P^{ef} and the correct average would, because of the latter's dependence on the quantity supplied, tend to produce a downward bias.[24] The same changes in omitted variables probably account for the high value of Wigley's total expenditure elasticity but hardly affect his price

(24) Compare 8.10.

elasticities because of the ratio form of the price variable and use of average prices for both electricity and gas.[25]

The estimated β-elasticities given in Table 6.9.1, when compared with the elasticities from (6.13.1) computed at the same value of unrestricted demand, are consistent with plausible Z effects, except for the price of gas: the comparison implies positive Z effects from increases in real total consumption expenditure $(\bar{E})$ or the price of solid fuel $(\bar{P}^c)$, and negative ones from increases in the price of electricity or gas availability (G); for all of them the elasticity due to the Z effect is larger absolutely than the estimated β-elasticity. For gas availability, the Z effect is, plausibly, so large and negative that it outweighs the positive Z-U effect of a fall in ownership.

The availability of gas was omitted from (6.13.2) because the method of deriving its functional relationship to quantity demanded by aggregation, as used in 3.8, leads to an intractable equation when loglinearity is assumed. As it is positively correlated with total expenditure, the effect is to bias downward the total expenditure coefficient, but the other coefficients should not be materially affected. Allowing for this and the change of dependent variable from unrestricted to total quantity demanded, the results with (6.13.1) and (6.13.2) are very similar when compared at the sample means of the dependent variables to minimize the effects of the change in functional form. The rapid increase in off-peak demand after 1965/66 and its high temperature sensitivity[26] account for the variations in the coefficient of temperature (W).

Because of the specification errors present in our two equations and those in Wigley (1968) and Stone (1954),[27] no very certain conclusions can be drawn from a comparison. The elasticities of demand with respect to prices obtained by Wigley for 1955-65 and Stone for 1920-38 appear not to be inconsistent with our estimated β-elasticities. Compared with before 1938, demand from all types of appliances together appears to have varied more since 1955 with changes in temperature between seasons and to have been more income-elastic.

(25) See 2.10.
(26) See Table 7.19.2.
(27) See 2.10 and 2.2 respectively.

CHAPTER 7

AN ANALYSIS OF DEMAND
ON OFF-PEAK TARIFFS

7.1 Introduction

As a consequence of the modification to the simple model of Chapter 1 introducing disaggregation over time, the analysis was divided into two parts, one concerned with demand on unrestricted tariffs, the other with demand on off-peak tariffs.[1] It is to the latter that we now turn.

We define as *off-peak tariffs* those tariffs variously called Restricted Hour and Off-Peak, offered by all Area Boards, giving a rate applicable only to consumption by specified appliances during restricted hours, and the night part of the Eastern Electricity Board's "night and day" tariff, applicable to consumption by all of a consumer's appliances.

The off-peak hours consisted of up to about twelve hours at night and about three hours in the afternoon. Area Boards usually offered several tariffs with different definitions of the off-peak period. We shall abbreviate "demand on off-peak tariffs" to *off-peak demand*; as noted in 1.8, this does not include all demand during the off-peak hours.

Before 1959 there was virtually no consumption on off-peak tariffs. In the late fifties and early sixties new types of appliances came on to the market which were capable of meeting a demand for heat hours after they had stopped using electricity, and standard off-peak tariffs were introduced and published. Thereafter ownership of appliances and consumption on off-peak tariffs rose very rapidly, but at different rates in each Area, as shown by Table 7.20.1 below. The question is then, were the high rate of growth of demand and the variations in it entirely due to Z effects, or can some Z-U and U effects be detected as well?

The analysis follows the same lines as that in Chapters 3 to 6. In 7.2-7.8, the model is derived; an assumption is not discussed if the implications are the same as before. Most of the variables entering the off-peak demand equation are the same as those in the unrestricted equation; the additional calculations are described in 7.9-7.13. The minor changes from Chapter 5 in

(1) See 1.8.

the methods of estimation and testing are set out in 7.14. Finally the results are analysed as in Chapter 6 (7.15-7.23).

7.2 The demand equation for an individual consumer

Analogous to the identity (3.2.1), the quantity demanded on off-peak tariffs by an individual consumer, h, due to use of off-peak appliances of type a, on day d of month m, may be written

$$Q_{dm}^{*\,oah} \equiv (Alur)_{dm}^{oah} \tag{7.2.1}$$

The superscript "o" denoting off-peak will, for the rest of this chapter, be omitted where no ambiguity arises. Assumptions made in earlier chapters are to be understood as if stated with the appropriate variables superscripted "o".

The types of off-peak appliances are assigned to two classes: class 1— space-heaters; class 2—water-heaters. The number of cookers and lighting and sundry appliances on off-peak tariffs was negligible. This classification should satisfy the independence condition of assumption 3.3.1. There were several distinct main types of space-heaters and water-heaters. Although some data are available on ownership of each type, they have had to be aggregated to overcome the estimation problems. Thus, as the classes probably do not satisfy the homogeneity condition of assumption 3.3.1, the changes in the proportions of each type in the total stock may produce Z-U effects as discussed in 3.6.

Assumption 3.3.2 alternative (i) is made for both classes, whereby ownership Z^{oah} and utilization U^{oah} are independent functions when defined as

$$Z^{oah} \equiv (Al)^{oah} \text{ the installed load}$$
and
$$U^{oah} \equiv (ur)^{oah} \tag{7.2.2}$$

There were large variations in the installed load of space-heaters per consumer having, l^{o1}, between Areas and it rose in England and Wales over time. The inter-Area variations in the installed load of water-heaters, l^{o2}, were smaller; it is not known whether this changed over time. Our hypothesis is that an increase was due to an increase in the demand for heat or, in the case of space-heaters, in the proportion of the demand for heat in a particular room met by off-peak heaters or in the number of rooms heated by them. Given the nature of the appliances, a higher load could not normally be used to meet the same demand for heat more quickly or more conveniently, as required by the alternative version of assumption 3.3.2. If, however, the increment in the load was used less intensively than existing heaters of the same type or raised the proportion in the total stock of a type with a relatively low utilization per kilowatt, the assumption is not valid and Z-U effects will result.

Making assumption 3.3.3 that utilization is a linear function of N variables X_j^h to be specified later and a random error term; and assumption 3.3.4 on the predeterminateness of the ownership variables, the recorded off-peak demand of consumer h, assumed to differ from actual demand by a random error ε^{*h}, is

$$Q^{oh} = \sum_{a=1}^{2} Z^{oah} (\sum_j \beta_{j,a}^h X_j^h + \varepsilon^{ah}) + \varepsilon^{*h} \qquad (7.2.3)$$

Assumption 3.3.4 should be much more nearly valid here than in the unrestricted analysis: as there, the true values of the ownership variables should be very nearly independent of the errors in the equation; the measurement errors are here much smaller as no interpolated values are used and the data appear to be much more accurate.[2]

7.3. Aggregation over consumers

We aggregate over all consumers, C—not just off-peak ones. Assumption 3.4.1 on the distribution of the $\beta_{j,a}^h$ is made here; this will be relaxed in 7.6 below to take account of variations in the availability of a gas supply.

The data again do not enable us to improve on assumption 3.4.2 that the weighted and unweighted means of the X_j^h are equal. Now even at the end of the period, in no Area in England and Wales were more than 8% of consumers on off-peak tariffs. Very large and varying differences between the weighted and unweighted means are therefore possible: if changes in a variable had large Z effects, the use of the unweighted mean will probably result in any U effects going unexplained.

Defining aggregate demand as

$$Q^o \equiv \sum_h Q^{oh} \qquad (7.3.1)$$

we obtain

$$\frac{Q^o}{C} = \sum_a Z^{oa} \sum_j \beta_{j,a}^* X_j^* + \varepsilon \qquad (7.3.2)$$

$$= \sum_a Z^{oa} (U^{oa} - \varepsilon^a) + \varepsilon \qquad (7.3.3)$$

where the terms on the right-hand side are means defined analogously to those in 3.4.

7.4. Variables in the utilization functions

Using the same postulates and empirical evidence from which our hypothesis about the utilization of heaters on unrestricted tariffs was derived, the hypothesis may be generalized and applied to the case of off-peak heaters. Thus we hypothesize that:

(2) See 7.10 below.

(1) the main determinants of the utilization of a given stock of off-peak space-heaters and water-heaters were temperature and the current and lagged values of mean total consumption expenditure per household, the price of off-peak electricity, the prices of its main substitutes given the stock of appliances—unrestricted electricity, solid fuel and gas—and the prices of all other commodities combined in a single index;

(2) utilization was a non-linear function, X_3^o, of temperature to be specified later;

(3) utilization was linear in the logarithms of total consumption and the prices; the lag distribution of total consumption expenditure was the simple 16-quarter one defined in 4.19, and of the prices one of the three alternatives in 4.19—these will be compared on the basis of regression results as before.

Unrestricted electricity is included here as a main substitute although off-peak electricity was not so included in the unrestricted equation because the proportion of off-peak consumers with unrestricted heaters was always much higher than the proportion of (unrestricted) consumers with off-peak heaters, so that a larger U effect on the average utilization might be expected. Also the proportion of off-peak consumers with unrestricted heaters was more nearly constant so that the complication of allowing for the dependence of the size of the U effect on the ownership level of the substitute appliances is unnecessary, whereas it would have been unsatisfactory to assume that the off-peak price coefficient in the unrestricted equation was constant.

The final rate P^{ef} is used as the best approximation available to the average marginal price of unrestricted electricity to off-peak consumers: only observations at the end of the period are used in this off-peak analysis, by which time the proportion of all consumers with any other marginal rate was small. The proportion of off-peak consumers was probably even smaller because of the positive correlation between ownership of appliances on off-peak tariffs and ownership of appliances on unrestricted tariffs.

We therefore define the set of variables entering the off-peak utilization functions as follows:

$$X_2^h \equiv 1 \tag{7.4.1}$$

$$X_3^h \equiv X_3^o(W^h) \tag{7.4.2}$$

$$X_7^h \equiv \text{lag}_0 \ln (\bar{E}/\bar{\pi})^h \tag{7.4.3}$$

$$X_8^h \equiv \text{lag} \ln (P^{ef}/\bar{\pi})^h \tag{7.4.4}$$

$$X_9^h \equiv \text{lag} \ln (\bar{P}^c/\bar{\pi})^h \tag{7.4.5}$$

$$X_{10}^h \equiv \text{lag} \ln (\bar{P}^g/\bar{\pi})^h \tag{7.4.6}$$

$$X_{80}^h \equiv \text{lag} \ln (P^{eo}/\bar{\pi})^h \tag{7.4.7}$$

where P^{eo} is the price of electricity on off-peak tariffs and the lag operator acts on it exactly as defined for the other price variables.

Assumption 3.5.1 on the distribution of these exogenous variables is made, and means X_j and corresponding coefficients $\beta_{j,a}$ defined as in 3.5. Hence

$$\frac{Q^o}{C}=\sum_{a=1}^{2} Z^{oa} \sum_{j=2, 3, 7, \ldots, 10, 80}\beta_{j,a}X_j \quad + \varepsilon \qquad (7.4.8)$$

7.5 Modifications to allow for the Z-U effects

Z-U effects may be expected as a result of invalidity of the assumptions already referred to and of assumption 3.5.2 on the completeness of the list of exogenous variables. In view of the large variations between Areas, and rapid growth, in the ownership of off-peak appliances, these may well be large. To allow for the average effect of rises in ownership from all causes being non-zero, we include in each utilization function

$$X_1^{(a)}=\frac{1}{Z^{oa}} \qquad (Z^{oa}>0) \qquad (7.5.1)$$

with coefficient $\beta_{1,a}$.

Only observations after 1964II are used in this analysis. Over the range of positive values of Z^{oa} experienced in this period, this should be a satisfactory approximation to the true specification, although obviously wrong at the lower values of Z^{oa} experienced earlier.

7.6 Modification to allow for variations in the availability of a gas supply

Allowance for variations in the availability of a gas supply implies as in the unrestricted analysis relaxation of assumption 3.4.1 on the distribution of the $\beta_{j,a}^h$. As before, it is assumed that the population of consumers may be dichotomized into those with and those without a gas supply and assumption 3.8.1 is made on the difference between the mean coefficients $\beta_{j,a}^G$ and $\beta_{j,a}^{NG}$ of the two groups. An expression analogous to (3.8.1) is then obtained in which the terms Z^{oaNG}/Z^{oa} and Z^{oaG}/Z^a appear, where Z^{oaNG} and Z^{oaG} are the ownership levels amongst those without and those with a gas supply respectively.

The availability of a gas supply is poorly correlated between Areas with both the ownership of off-peak space-heaters and the ownership of off-peak water heaters: $r(Z^{o1}, G)=-0.28; r(Z^{o2}, G)=-0.05$. This could mean that assumption 3.8.2(ii), implying lack of substitutability between gas and off-peak electrical appliances, is nearly valid, or that there is a relationship between gas availability and off-peak ownership *ceteris paribus* which is obscured by correlation with other variables. In the latter case, however, as long as the correlation with the other variables is present within Areas, Z^{oaNG}/Z^{oa} and Z^{oaG}/Z^{oa} would still be nearly constant. We therefore make the same modification as before—replacing X_{10} as defined in (7.4.6) above by

$$X_{10} \equiv G \text{ lag ln } (\bar{P}^g/\bar{\pi}) \tag{7.6.1}$$

and adding to the utilization functions

$$X_{11} \equiv G \tag{7.6.2}$$

7.7 The final form

As a result of the modifications introduced in the preceding two sections, the off-peak demand equation for day d of month m is

$$\frac{Q^o}{C} = \beta_1 + \sum_{a=1}^{2} Z^{oa} \sum_{j=2, 3, 7, \ldots, 11, 80} \beta_{j,a} X_j + \varepsilon \tag{7.7.1}$$

where $\beta_1 \equiv \beta_{1,1} + \beta_{1,2}$.

Assumption 3.10.1 on homoscedasticity of ε is made again. If the individual errors ε^{ah} and ε^{*h} were homoscedastic—so that the variance of ε were given by an expression analogous to (3.10.1)—there is a stronger presumption here than in the unrestricted analysis that the variance of the mean error ε rose over time, because of the more rapid rise in ownership. However, because households were acquiring types of appliances that were new to them and necessitated new habits of utilization, any one household's utilization probably had a relatively large random element in it at first. Secondly, if the speed of adjustment to a new appliance stock equilibrium was poorly correlated with the included exogenous variables and the omitted variables that contribute most to the error variance, the difference between the mean values of these variables for new and existing off-peak consumers may have varied considerably from quarter to quarter. Given the decline in the proportional rate of growth of ownership, both these factors would tend to produce a decline in the variance of the mean error ε.

Inclusion of the price of off-peak electricity presents no new identification problem. In the short run as calculated it is independent of the quantity supplied. Similar arguments to those advanced in 3.11 when discussing the price of unrestricted electricity lead us to expect only a weak supply relationship in the long run between the off-peak price and current consumption on off-peak tariffs.

7.8 Aggregation over time

The series for Q^o—as for Q^u—give units billed in each quarter.

Let Q^o_q be units billed in quarter q,

       M^{om}_q the proportion of month m's off-peak consumption read in quarter q.

Then

$$Q_q^o = \sum_{m=3q-5}^{3q} M_q^{om} \sum_{d=1}^{D^m} Q_{dm}^o \qquad (7.8.1)$$

Assumptions 4.4.1 that Q^o and C, the number of consumers, are un-correlated within quarters and 4.4.2 that Z^{oa} and X_j are uncorrelated within months are made. Defining moving-quarter means Z_q^{oa}, X_{qj}, and ε_q as in 4.4 and mean off-peak quantity demanded per consumer per day in moving-quarter q as

$$Y_q^o \equiv \frac{Q_q^o}{C_q \sum_m M_q^{om} D^m} \qquad (7.8.2)$$

and aggregating the demand function (7.7.1) over time, we then obtain

$$Y_q^o = \beta_1 + \sum_{q=1}^{2} Z_q^{oa} \sum_{j=2, 3, 7 \ldots, 11, 80} \beta_{j,a} X_{qj} + \varepsilon_q \qquad (7.8.3)$$

This is the final form to be estimated. The error variances and covariances are as given by (4.5.1).

7.9 Calculation of the variables

Lacking any information on the reading of off-peak meters, we assume that the organization is the same as for other meters and hence that

$$M_q^{om} = M_q^m \qquad \text{all } m, q$$

where M_q^m is as calculated in the unrestricted analysis. Values calculated using the simpler assumptions 4.6.1 and 4.6.2 were tried and found to give less satisfactory results—a marginally worse fit and more seasonal pattern in the residuals.

There are no data on the exogenous variables from which the differences between the means for all consumers and for off-peak ones alone can be estimated. Thus for variables which appeared in the unrestricted analysis, the same series are used as defined earlier; the only ones remaining to be calculated are for off-peak ownership—Z^{o1}, Z^{o2}—, quantity demanded—Q^o—, and price—P^{eo}—, and the transformed temperature variable $X_3^o(W)$. In the following sections, the calculations for Areas in England and Wales are described. As before some data are available on the South of Scotland and are used for predictive testing; the calculations were similar.

7.10 Ownership of off-peak space-heaters and water-heaters—Z^{o1}, Z^{o2}

Off-peak appliance class 1 is defined as those appliances of the storage, floor-warming and ducted air "indirect-acting" types connected to off-peak meters; and class 2 as all water-heaters connected to off-peak meters.

The series for ownership of off-peak space-heaters, Z^{o1}, and ownership of off-peak water-heaters, Z^{o2}, were obtained from the Electricity Council's records of "installed load on restricted hour tariffs—all retailer sales with adjustments for disconnections: total load connected at 30 June, 30 September, 31 December, 31 March each year, commencing June 1964". For all Areas for at least one type of appliance, we were only able to obtain series covering part of the period from June 1964 onwards, or for the total, as distinct from domestic load. Deriving figures for Z^{o1} and Z^{o2} only for those quarters for which we know all the major elements of the total installed load in both classes, we obtain 106 observations. This number is sufficient to give a reliable estimate of the demand equation; additional figures obtained simply by trend extrapolation could be subject to large proportional errors given the large variations in growth rates. Only these 106 observations, the ones listed in Table 7.10.1, are used therefore in estimating the demand equation.

TABLE 7.10.1. *The observations from which the off-peak equation was estimated*

Area	Quarters	Area	Quarters
London	1964ɪᴠ-68ɪ	Midlands	1967ɪɪɪ-68ɪ
South Eastern	1967ɪɪ-68ɪ	South Wales	1965ɪɪɪ-68ɪ
Southern	1965ɪɪ-68ɪ	Merseyside and North Wales	1967ɪɪɪ-68ɪ
South Western	1965ɪɪ-68ɪ	Yorkshire	1964ɪᴠ-68ɪ
Eastern	None	North Eastern	1964ɪᴠ-68ɪ
East Midlands	1964ɪɪɪ-67ɪɪ	North Western	1966ɪᴠ-68ɪ

Total number of observations = 106

These series should be much more reliable than those for the unrestricted classes because they are not based on sample surveys and include no interpolated values. There is a small inaccuracy in the space-heating series because class 1 as defined does not include all space-heaters on off-peak tariffs: the error never exceeded 3% and was for most observations negligible.

The extended series running from 1955ɪɪ to 1968ɪ which were needed in the unrestricted analysis were calculated after estimation of the off-peak equation, as it was then possible to use a more reliable method of extrapolation: both Z^{o1} and Z^{o2} were taken to be 0·000 kilowatts until 1958ɪᴠ; values between 1958ɪᴠ and the first quarter given in Table 7.10.1 for any particular Area were then interpolated using the actual consumption figures and regression estimates of utilization.

7.11 The quantity variable—Q^o

For all the 106 observations, Q^o, the off-peak consumption of specified appliances separately metered, was available directly from records on the whole population of consumers. Thus the series for the dependent variable Y^o are very accurate.

7.12 The price of off-peak electricity—P^{eo}

Area Boards typically offered several off-peak tariffs differing in respect of the times at which an off-peak supply was available. The four main types gave about eight or twelve hours at night with or without three hours during the afternoon; they consisted of a fixed charge and a single marginal rate, which was higher the greater the number of hours classified as off-peak. The only information we have on the proportion of consumers on each tariff is that the one giving eight hours at night and three hours in the afternoon was the most popular in the South West. From 1965 onwards, all Boards had such a tariff.

P^{eo} has therefore been calculated as the minimum marginal rate for a supply for at least eight hours at night and three hours during the day. As the tariffs are directly available, the only error in the variable arises from the difference between P^{eo} and the true determinant of aggregate U effects—the weighted average of the marginal rates. This difference must have been nearly constant in all Areas and quarters in which all four types of tariff were available unless the proportion of consumers on each type varied greatly, as the differentials between the rates varied little. In the few instances in which a rate for more than eleven hours has been used because no "eight-plus-three hours" tariff was offered, the difference would still have been about the same as long as the proportion of off-peak consumers using the "eight-plus-three hours" tariff when available was high.

7.13 The temperature variable—W^o

In a daily off-peak demand function, the weights attached to temperatures at different hours of the day would not be the same as those in a daily unrestricted demand function, given that most off-peak space-heaters were indirect-acting. But over a whole month the average of the relevant temperatures should be about the same. The same series W_m, as defined in 4.20, is therefore used.

On the basis of the same arguments as used in 4.20, it is postulated that off-peak demand was insensitive to temperature changes above some threshold value τ^o, not necessarily equal to τ^u; that, below this, temperature response was non-linear; and that the semi-log form is a good approximation over the observed range. Thus X_3^o is defined as

$$X_3^o \equiv \ln W^o \tag{7.13.1}$$

where
$$W_m^o = \min (W_m, \hat{\tau}^o)$$

and $\hat{\tau}^o$ is a regression estimate of τ^o.

7.14. Estimation and testing: methods

With appropriate adjustments to the ranges of subscripts, most of Chapter 5 carries over to estimation of the off-peak equation.

In the analysis of the pattern of residuals,[3] the matrix **D** now contains one fewer Area dummies as there are no Eastern Area observations. The seasonal dummies with means removed are now not orthogonal to the Area ones, but the correlation is low enough to mean that $F\{c_2^o, c_3^o, c_4^o\}$ can still be used as a measure of seasonality. The series for individual Areas are now not long enough to justify calculation of $\hat{\rho}_a$ or d_B.

If assumptions 5.3.1-5.3.4 and 3.10.1 justifying the use of Generalized Least Squares were valid, the critical values of F- and t-test statistics would be slightly higher than in the unrestricted analysis: 3·93 to 3·95 for a one-tailed F-test of a single coefficient at the 5% level, depending on the number of independent variables included, and 1·98 to 1·99 for the corresponding t-test. The remarks made in 5.10 qualifying the inferences that can be drawn given the probable violation of the assumptions apply here with one exception: as the errors in the ownership variables are here much smaller, values of test statistics only slightly greater than the critical values given do, if no serial correlation is present, indicate more certainly rejection of the hypothesis under test.

7.15 The demand equation to be estimated

Bringing together the developments of the preceding sections, the equation for demand on off-peak tariffs, previously given as (7.8.3), may be written as

$$Y^o = \beta_1 + \sum_{a=1}^{2} Z^{oa} [\beta_{2,a} + \beta_{3,a} \ln W^o + \beta_{7,a} \text{ lag}_0 \ln (\bar{E}/\bar{\pi})$$
$$+ \beta_{8,a} \text{ lag} \ln (P^{ef}/\bar{\pi}) + \beta_{9,a} \text{ lag} \ln (\bar{P}^c/\bar{\pi})$$
$$+ \beta_{10,a} G \text{ lag} \ln (\bar{P}^g/\bar{\pi}) + \beta_{11,a} G$$
$$+ \beta_{80,a} \text{ lag} \ln (P^{eo}/\bar{\pi})] + \varepsilon \qquad (7.15.1)$$

where, to recapitulate definitions given fully in the preceding sections and Chapter 4,

Y^o is the quantity demanded of electricity on off-peak tariffs per consumer per day

$a=1$ denotes space-heaters

$a=2$ water-heaters

Z^{oa} is the off-peak appliance ownership level

W^o temperature, the lower of actual temperature and an estimated threshold, $\hat{\tau}^o$

$\bar{E}$ total consumption expenditure per household

$\bar{\pi}$ an all-commodities price index

(3) See 5.6.

P^{ef} the final rate on the unrestricted electricity tariff
$\bar{P}^c$ the price of solid fuel
$\bar{P}^g$ the price of gas
G the proportion of electricity consumers having a gas supply available
P^{eo} the price of off-peak electricity
lag   an operator giving a weighted average of current and past values of the operand.

The interpretation of the coefficients is similar to that of the coefficients in the equation for unrestricted demand discussed in 6.1: *ex hypothesi* the constant β_1 is the sum of two constants measuring the average Z-U effects of changes in the stocks of each of the two classes of appliances. A coefficient $\beta_{j,a}$ measures the sum of the U effect and Z-U effect of changes in the corresponding variable on the utilization of appliances in class a, less the average Z-U effect associated with changes in the stock of appliances in this class from all causes.

7.16 Estimation of the temperature threshold parameter, τ^o

As a first step towards estimating the demand equation (7.15.1), the temperature threshold parameter τ^o was estimated. Temperature is virtually uncorrelated with the other exogenous variables so that, in the interests of computational efficiency, the parameter was estimated using the equation

$$Y^o = b_1 + b_{2,1}Z^{o1} + b_{2,2}Z^{o2} + b_{3,1}Z^{o1}\ln W^o + b_{3,2}Z^{o2}\ln W^o + \varepsilon$$

$$(7.16.1)$$

This was estimated by Generalized Least Squares for values of $\hat{\tau}^o$ at intervals of $0 \cdot 5$ °C.[4] The results in Table 7.16.1 indicate the general pattern.

TABLE 7.16.1. *Off-peak demand: results with different estimates of the temperature threshold*
Equation estimated: (7.16.1)

$\hat{\tau}^o$ °C	∞ [1]	15·0	14·0	11·0
$b_{3,1}$	$-13\cdot5$	$-14\cdot1$	$-14\cdot8$	$-19\cdot0$
$b_{3,2}$	$-11\cdot8$	$-12\cdot3$	$-12\cdot4$	$-15\cdot0$
$F\{b_{3,1}^o\}$	4994	5701	5306	2229
$F\{b_{3,2}^o\}$	19·57	22·95	20·12	7·83
R^2	0·988	0·990	0·989	0·975
$t\{\beta\}$	$-0\cdot81$	$-0\cdot82$	$-1\cdot17$	$-1\cdot06$
$F\{c_5^o, \ldots, c_{14}^o\}$	3·22	2·87	2·36	1·50
$F\{c_2^o, c_3^o, c_4^o\}$	9·21	4·73	1·92	12·10

(1) These results apply for all $\hat{\tau}^o > 18\cdot5$.

(4) Greater computational difficulties arising from the uneven length of the eleven time-series precluded the use of intervals as fine as those in the unrestricted analysis.

The conclusion drawn is that the best estimate of the mean quarterly temperature above which mean quarterly off-peak demand was virtually constant is 14·0 °C when temperature is measured as the mean at 09.00 and 21.00 hours. Hence $\hat{\tau}^o = 14·0$ °C has been used in further work. This minimizes the seasonal pattern of the residuals and there are no contraindications from other statistics. The total sum of squared residuals is only very slightly above the minimum attained at $\hat{\tau}^o = 15·0$.

Comparison of the results with $\hat{\tau}^o = \infty$ and $\hat{\tau}^o = 14·0$ shows that, as in the unrestricted analysis, it is important to allow for non-linearity. The hypothesis that the threshold is the same in all Areas is not refuted by the evidence on the Area pattern of the residuals.

7.17 Selection of the lag distribution

The demand equation (7.15.1) contains seventeen coefficients. There is no prior information which leads us to expect the coefficient of any exogenous variable to be closer to zero in one utilization function than the other. The equation is therefore estimated first without any constraints. Although the correlations between the explanatory variables are not so high as in the unrestricted analysis, the reliability of individual coefficients is still less than satisfactory. The results are used therefore only to select the lag distribution and to derive estimates $\bar{T}^{oa}$ of the identifiable part of each utilization function —evaluated at the mean of the 106 observations—analogous to $\bar{T}^a$ defined in 6.4, as a basis for imposing constraints.

There is scarcely any difference between the results with each of the three lag distributions. Whichever weights are used, the lagged variables collectively make no significant contribution to the explanation of variations in demand: with weights w_1, the twelve variables in total expenditure, prices and gas availability have $F = 1·87$. The complete results with this set of weights are given in Table 7.17.1.

Weights w_1 from a lognormal distribution with mode 1·5, implying that the rate of adjustment is a maximum one-quarter after a price change, have been used in further work as there is some evidence that the U effects are best explained with these weights. The autocorrelation coefficient, $\hat{\rho}$, is negative for all the lag distributions. The values of $\hat{\rho}$ and $t\{\hat{\rho}\}$ are closest to zero for w_1. Failure to explain Z-U effects would tend to give rise to positive serial correlation since they are long-run phenomena, but failure to explain U effects is more likely to produce negative serial correlation because the weights allocated to the current and immediately preceding quarters are wrongly distributed. Hence if the negative serial correlation observed is due directly to misspecification of the lags, U effects would appear to be best explained when weights w_1 are used.

TABLE 7.17.1. *Off-peak demand: results*

Variable k	b_k	$\hat{\sigma}_k$	$F\{b_k\}$	$F\{b_k^0\}$
1	0·0490	0·0407	1·45	11476
Z^{o1}	104	19·2	29·00	4116
$Z^{o1} \ln W^o$	−15·1	0·458	1089	5831
$Z^{o1} \lag_0 \ln (\bar{E}/\bar{\pi})$	−6·12	3·80	2·59	3·60
$Z^{o1} \lag_1 \ln (P^{eo}/\bar{\pi})$	−0·494	1·66	0·09	15·34
$Z^{o1} \lag_1 \ln (\bar{P}^c/\bar{\pi})$	5·46	2·41	5·12	36·16
$Z^{o1} \lag_1 \ln (P^{ef}/\bar{\pi})$	−1·11	2·71	0·17	6·10
$Z^{o1} G \lag_1 \ln (\bar{P}^g/\bar{\pi})$	−0·836	1·43	0·34	22·73
$Z^{o1} G$	6·74	4·68	2·07	1·25
Z^{o2}	27·5	89·5	0·09	44·09
$Z^{o2} \ln W^o$	−12·9	3·29	15·43	20·33
$Z^{o2} \lag_0 \ln (\bar{E}/\bar{\pi})$	42·8	20·2	4·51	0·17
$Z^{o2} \lag_1 \ln (P^{eo}/\bar{\pi})$	−12·4	17·2	0·53	0·00
$Z^{o2} \lag_1 \ln (\bar{P}^c/\bar{\pi})$	−37·9	23·5	2·60	3·47
$Z^{o2} \lag_1 \ln (P^{ef}/\bar{\pi})$	−3·04	22·3	0·02	0·76
$Z^{o2} G \lag_1 \ln (\bar{P}^g/\bar{\pi})$	16·0	12·4	1·66	0·00
$Z^{o2} G$	−74·8	34·2	4·78	4·78

$$R^2 = 0·991 \qquad\qquad F\{R^2\} = 631·6$$

$\bar{T}^{o1} = 2·90$
$\bar{T}^{o2} = 4·76$

7.18 Constraints on elasticities

The relationship postulated between the coefficients in the two utilization functions is that the elasticities are the same at the sample means:

$$\frac{\beta_{j,1}}{\bar{T}^{o1}} = \frac{\beta_{j,2}}{\bar{T}^{o2}} \qquad j = 7, 8, 9, 10, 11, 80 \qquad (7.18.1)$$

where $\bar{T}^{oa}$ is the estimate of the identifiable part of the utilization function given in Table 7.17.1. This form is chosen because, for the same reasons as given in 6.5, it seems likely to be the most realistic of the feasible alternatives.

7.19 The constrained demand equation

On incorporating these constraints in (7.15.1) and using the results of 7.16-7.17, the demand equation becomes

$$Y^o = \beta_1 + \beta_{2,1}Z^{o1} + \beta_{2,2}Z^{o2} + \beta_{3,1}Z^{o1} \ln W^o + \beta_{3,2}Z^{o2} \ln W^o$$
$$+ \beta_{7,1}K^0 \lag_0 \ln (\bar{E}/\bar{\pi}) + \beta_{80,1}K^0 \lag_1 \ln (P^{eo}/\bar{\pi})$$
$$+ \beta_{9,1}K^0 \lag_1 \ln (\bar{P}^c/\bar{\pi}) + \beta_{8,1} K^0 \lag_1 \ln (P^{ef}/\bar{\pi})$$
$$+ \beta_{10,1} K^0 G \lag_1 \ln (\bar{P}^g/\bar{\pi}) + \beta_{11,1}K^0 G$$
$$+ \varepsilon \qquad\qquad (7.19.1)$$

with 17 variables and lag weights w_1

Dummy variable k	c_k	$F\{c_k\}$	$F\{c_k\}$
1	−0·0688	0·12	0·01
Moving-quarter:			
I	0·239	2·36	0·09
II	0·366	4·83	3·44
III	0·252	2·41	2·53
Area:			
London	−0·169	0·53	0·07
South Eastern	0·160	0·22	0·95
Southern	−0·205	0·71	0·18
East Midlands	0·0231	0·01	0·90
Midlands	−0·366	0·91	0·39
South Wales	−0·402	2·63	2·53
Merseyside and North Wales	0·0192	0·00	0·14
Yorkshire	−0·163	0·49	0·25
North Eastern	−0·00504	0·00	0·28
North Western	−0·362	1·40	1·40

$$F\{c_2^o, c_3^o, c_4^o\} = 2 \cdot 02$$
$$F\{c_5^o, \ldots, c_{14}^o\} = 0 \cdot 71$$

Autocorrelation coefficient: $\hat{\rho} = -0 \cdot 0681$ $\quad t\{\hat{\rho}\} = -0 \cdot 60$

Heteroscedasticity coefficient: $\hat{\psi} = 0 \cdot 0934$ $\quad t\{\hat{\psi}\} = 1 \cdot 46$

where $$W_m^o = \min (W_m, 14 \cdot 0 \ ^{\circ}\text{C})$$

and

$$K^0 = Z^{o1} + \frac{\overline{T}^{o2}}{\overline{T}^{o1}} Z^{o2} \tag{7.19.2}$$

The results of estimating this by Generalized Least Squares are given in Table 7.19.1. The derived β-elasticities of utilization defined analogously to those in 6.9 are given in Table 7.19.2.

7.20 Goodness of fit

The results show that variations in off-peak demand were almost entirely accounted for by variations in appliance ownership, utilization remaining constant, and in natural conditions.

Table 7.20.1 shows the extent to which the equation explains variations in the off-peak component of the annual consumption and growth rate figures given in Table 1.3.1: it explains well the level of off-peak consumption at the end of the period, and explains the growth rate in the latter part of the period better than the unrestricted demand equation. This reflects the relative insignificance of the price and total consumption expenditure variables and the greater accuracy of the ownership ones.

TABLE 7.19.1. *Off-peak demand: results*

Variable k	b_k	$\hat{\sigma}_k$	$F\{b_k\}$	$F\{b_k^o\}$
1	0·0312	0·0327	0·91	11042
Z^{o1}	94·0	7·32	164·7	3961
Z^{o2}	74·3	20·3	13·46	88·5
$Z^{o1} \ln W^o$	$-15\cdot2$	0·413	1353	5631
$Z^{o2} \ln W^o$	$-12\cdot2$	2·70	20·47	21·35
$K^o \operatorname{lag}_0 \ln (\bar{E}/\bar{\pi})$	$-1\cdot23$	1·28	0·91	1·35
$K^o \operatorname{lag}_1 \ln (P^{eo}/\bar{\pi})$	$-1\cdot33$	0·910	2·13	0·69
$K^o \operatorname{lag}_1 \ln (\bar{P}^c/\bar{\pi})$	1·84	0·802	5·27	7·26
$K_o \operatorname{lag}_1 \ln (P^{ef}/\bar{\pi})$	$-1\cdot42$	1·21	1·38	0·69
$K^o G \operatorname{lag}_1 \ln (\bar{P}^g/\bar{\pi})$	1·04	0·744	1·94	1·56
$K^o G$	$-1\cdot32$	1·66	0·64	0·64

$$R^2 = 0\cdot990 \qquad\qquad F\{R^2\} = 971\cdot4$$

Brief definitions of the variables appearing were given in 7.15, full definitions in Chapter 4 and 7.10-7.13. K^o is the weighted average of the two ownership levels of off-peak space-heaters and water-heaters defined by (7.19.2).

TABLE 7.19.2. *Off-peak demand: β-elasticities estimated from equation* $(7.19.1)^{(1)}$

Variable	β-elasticity
$\bar{E}$	$-0\cdot40$
P^{eo}	$-0\cdot44$
$\bar{P}^c$	0·60
P^{ef}	$-0\cdot47$
$\bar{P}^g$	0·25
G	0·40
W^o	$-4\cdot6$

(1) Calculated at the means of the 106 observations.

with the constrained equation (7.19.1)

Dummy variable k	c_k	$F\{c_k\}$	$F\{c^o\}$
1	0·0311	0·02	0·01
Moving-quarter:			
I	0·238	2·30	0·09
II	0·366	4·76	2·99
III	0·261	2·55	2·58
Area:			
London	− 0·248	1·11	0·01
South Eastern	− 0·350	1·03	0·17
Southern	− 0·254	1·08	0·03
East Midlands	− 0·00227	0·00	1·88
Midlands	0·0766	0·04	1·02
South Wales	− 0·811	10·54	10·67
Merseyside and North Wales	− 0·196	0·32	0·02
Yorkshire	− 0·336	2·04	1·80
North Eastern	− 0·0301	0·02	0·16
North Western	− 0·349	1·36	1·36

$$F\{c_2^o, c_3^o, c_4^o\} = 1\cdot89$$
$$F\{c_5^o, \ldots, c_{14}^o\} = 1\cdot71$$

Autocorrelation coefficient: $\hat{\rho} = - 0\cdot0624$ $t\{\hat{\rho}\} = - 0\cdot54$

Heteroscedasticity coefficient: $\hat{\psi} = 0\cdot0746$ $t\{\hat{\psi}\} = 0\cdot83$

TABLE 7.20.1. *Consumption on off-peak tariffs by Area*

Actual and predicted[1] figures (predicted figures are in brackets)

Electricity Board	Off-peak units billed per domestic consumer	
	Growth rate '64/65 to '67/68 % p.a.	1967/68 kWh
London	39·8 (33·0)	174 (171)
South Eastern	46·7 (46·8)	361 (364)
Southern	45·7 (45·8)	503 (509)
South Western	43·7 (46·7)	473 (480)
Eastern	31·8 (34·4)	492 (475)
East Midlands	47·1 (41·0)	287 (263)
Midlands	35·1 (33·9)	422 (410)
South Wales	35·8 (37·8)	236 (269)
Merseyside and North Wales	52·8 (47·7)	178 (176)
Yorkshire	44·3 (42·0)	377 (385)
North Eastern	49·7 (46·7)	234 (217)
North Western	41·9 (40·4)	349 (354)
England and Wales	41·2 (40·5)	352 (349)

(1) For quarters not included in the set of regression observations (see Table 7.10.1), the predicted figures are obtained using estimates of appliance ownership. These are derived by a procedure such that the prediction errors for the Eastern Electricity Board are biased towards zero. For other Areas, the bias is small enough to be ignored.

7.21 The pattern of the residuals

The weakness of the pattern of the residuals suggests that the errors come much closer than they did in the unrestricted analysis to satisfying the assumptions on randomness and homoscedasticity.

There is no significant serial correlation or heteroscedasticity. This provides support for the hypothesis of 7.7 that there was a relatively large random element in the utilization of new owners and a low correlation between the determinants of utilization and the speed of adjustment to a new stock equilibrium, so that the effect of rising ownership in increasing the variance of the mean error was cancelled out.

As a consequence of using a temperature threshold of 14·0 °C there is no significant seasonal pattern. The Area constants are also collectively insignificant.

7.22 The interpretation of the coefficients

The total expenditure, price, and gas availability coefficients are collectively insignificant—$F = 2·03$—and also individually so except the solid-fuel price coefficient. It may be that the use of unweighted means of the variables reduced their explanatory power,[5] but as the residual variance is so small, the *sum* of the U and Z-U effects due to changes in the weighted means could not have been large. That the U and Z-U effects were separately both large but of opposite signs is unlikely: owners of off-peak water-heaters would not normally have owned other water-heaters. Off-peak space-heaters were fixed and often in rooms where there was no other fixed heater so that the only possible substitute would be a portable direct-acting unrestricted heater; in rooms where there was another fixed heater this would in many cases have been because both were needed to meet demand at certain times. Given these limitations on the scope for substitution and that direct-acting space-heaters would probably not be regarded as close substitutes for the off-peak indirect-acting ones, the U effects must have been small. Hence, since the constant b_1 is insignificant, the Z-U effects could not have been large either. That this was so despite the large rise in ownership over time is explicable if this increase was mainly a movement towards a new stock equilibrium following the introduction of new types of appliances and publication of standard off-peak tariffs, subsequent shifts in the equilibrium due to changes in the exogenous variables being relatively small; and if the speed of adjustment of individual households to their new stock equilibrium was poorly correlated with the exogenous variables.

When interpreting an individual coefficient, it has to be remembered, as in the unrestricted analysis, that it, and the elasticity given in Table 7.19.2,

(5) See the discussion of assumption 3.4.2 in 7.3 above.

do not measure the total effect on the off-peak demand for electricity; and that it is quite possible for the sum, which *ex hypothesi* the coefficient measures, of the U effect and the deviation of the Z-U effect from the average to be of the opposite sign to the total effect.

The significant positive coefficient of the solid-fuel price, $\bar{P}^c$, does indicate the presence of some positive Z-U effects, since the elasticity of 0·60 is high to be due entirely to U effects. There are two possible explanations of this. The positive Z-U effects could be the effects of changes in the weighted mean of total consumption expenditure, $\bar{E}$: as in the equation for unrestricted demand, they could be taken up by the solid-fuel price coefficient because of the positive correlation between the price of solid fuel and the difference between the weighted and unweighted means of total consumption expenditure.[6] Alternatively the explanation could be that where the price of solid fuel was low, ownership of solid-fuel central heating was substituted for ownership of off-peak heating mostly among households which would have used off-peak heaters relatively intensively, because of correlation between the price of solid fuel and some omitted taste variable.

The negative temperature coefficient for water-heating shows that off-peak water-heaters unlike unrestricted ones were predominantly owned by households which used them all the year round so that mean utilization is a monotonic decreasing function of temperature.

The finding that off-peak demand in total is more sensitive than unrestricted demand to changes in the mean temperature W between 11·7 and 14·0 °C may partly be explained by this difference in the sensitivity of water-heating demand. The rest of the explanation could be that off-peak owners desire a relatively high room temperature—because ownership and desired temperature are both functions of, say, income—or that they experience a lower mean temperature than other consumers because of either the types of housing they occupy or their situation.

7.23 Estimates of utilization

The utilization of both classes of off-peak appliances was probably fairly close to the estimates $\hat{T}^{oa}$ of the identifiable part of each function. The two series for England and Wales, shown in Figure 7.23.1, display no marked trend. As the constant b_1 is positive, there may have been a slight downward trend in the utilization of one or both classes. On the evidence of the average levels of these series and those for individual Areas, it is unlikely that b_1 is small as a result of the contributions to it from each of the utilization functions, $b_{1,1}$ and $b_{1,2}$, being large and of opposite signs.

(6) See 4.17.

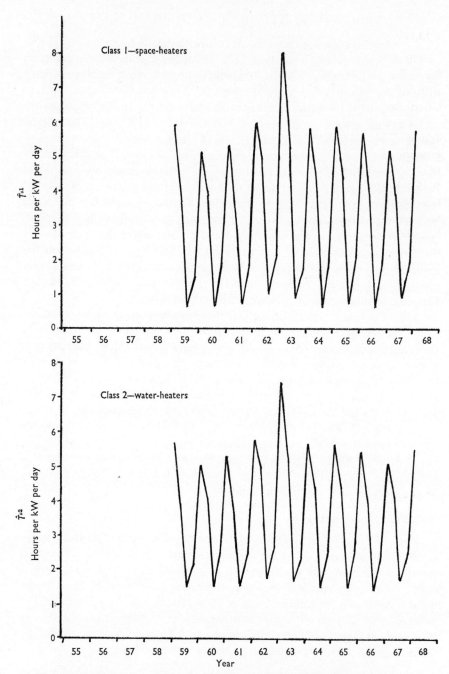

FIGURE 7.23.1. *The estimates of the identifiable parts of the off-peak utilization functions for England and Wales from the results with equation (7.19.1).*

7.24. Prediction

Off-peak demand, Y^o, in the South of Scotland Area has been predicted for the moving-quarters for which the actual ownership figures are known with the same reliability as the 106 observations on which the regression is based; the results are given in Table 7.24.1.

Comparison of the actual figures for the South of Scotland and England and Wales in Table 7.24.2 shows that the South of Scotland had the lower growth rate at the end of the period but a much higher level of annual off-peak consumption per consumer—higher than in any Area in England and Wales. The equation correctly predicts these rankings but overpredicts the level in the South of Scotland. This prediction error would probably have been even larger if the definition of the installed load of off-peak space-heaters had been extended to include types of space-heaters other than the three main ones, as the ownership of other types was much higher in the South of Scotland than in any other Area for which data are available.

TABLE 7.24.1. *Predictions of off-peak demand in the South of Scotland from equation (7.19.1)*

Predicted figures are given in brackets after the actual figures, with the standard errors underneath

| Moving-quarter | Kilowatt-hours per consumer per day | | | |
	II	III	IV	I
Year				
1964/65	(1)	0·42 (0·41) 0·10	(1)	(1)
1965/66	1·36 (1·75) 0·13	0·72 (0·67) 0·14	1·09 (1·25) 0·14	2·41 (2·86) 0·17
1966/67	1·90 (2·30) 0·17	0·94 (0·82) 0·18	1·36 (1·59) 0·19	3·06 (3·15) 0·21
1967/68	2·46 (2·93) 0·24	1·06 (1·13) 0·23	1·64 (1·92) 0·24	4·06 (4·52) 0·27

(1) Data not available for predictions.

TABLE 7.24.2. *Annual consumption on off-peak tariffs in the South of Scotland*

Predicted figures are given in brackets after the actual figures

Off-peak units billed per domestic consumer	Growth rate '65/66 to '67/68 % p.a.	1967/68 kWh
South of Scotland	28·2 (26·5)	841 (957)
England and Wales	36·7 (35·6)	352 (349)

That the growth rate is predicted fairly well favours the hypothesis[7] that some different cause of growth was at work in the South of Scotland such that mean utilization was lower in the prediction period and had been lower there at the ownership levels experienced in England and Wales, rather than the alternative hypothesis that the prediction error is largely due to extrapolation beyond the range of ownership levels experienced in England and Wales on the assumption of no Z-U effects.

(7) See 6.12.

CHAPTER 8

A VARIANT MODEL WITH A
TARIFF FUNCTION

8.1 Introduction

Returning to the analysis of unrestricted demand, the conclusion reached in our analysis in 4.14 was that the appropriate price of unrestricted electricity in the long-run equilibrium unrestricted demand function for an individual household is the marginal rate given by its tariff function at the equilibrium consumption level; and in the aggregate function an average of the marginal rates weighted for each utilization function U^a by the corresponding ownership values Z^{ah}.

Previously the final rate P^{ef} has been used because it was the marginal rate for most consumers and hence should be highly correlated with the correct variable, and because the model is then simplified, allowing examination of the effects of varying other parts of the specification. Now that this examination has been completed, a model can be developed in which an average of the various rates is used.

This model is more complex firstly because the proportion of consumers on each marginal rate varied so that the average is dependent on the quantity supplied. A second equation representing the aggregate of the tariff functions must therefore be included and estimated simultaneously. Secondly, we have previously allowed for households being out of equilibrium because they were slow to adjust to new price levels; now we need to allow for them also being out of equilibrium because (i) the actual values of the determinants of demand turned out to be different from their expectations at the beginning of the quarter; and (ii) possibly they did not adjust to the regular, and expected, seasonal variations in price brought about by changes in natural conditions.

The final form of the model therefore consists of a demand function and a price function related to the aggregate tariff function. To simplify notation, the third relationship, quantity demanded equals quantity supplied, will be substituted into the tariff function from the beginning.

The model is developed primarily in order to investigate the sensitivity of the results reported in Chapter 6 to the use of the final rate P^{ef}. We shall also (in 8.9) look at the bias introduced by using an average marginal rate without allowing for its dependence on the quantity supplied.

8.2 The demand equation

The demand equation used is the same as the constrained one in Chapter 6, (6.6.1), except that for the final rate P^{ef} is substituted P^{eu}, an average of the marginal rates to be defined:

$$
\begin{aligned}
Y^u = \beta_1 &+ \beta_{2,1}Z^1 + \beta_{2,2}Z^2 + \beta_{2,3}Z^3 + \beta_{2,4}Z^4 \\
&+ \beta_{3,1}Z^1 \ln W^u + \beta_{3,2}Z^2 \ln W^u + \beta_{4,4}Z^{41} \ln (L/\bar{L}) \\
&+ \beta_{5,1}Z^1Z^{o1} + \beta_{6,2}Z^2Z^{o2} \\
&+ \bar{\beta}_7 K^4 \, \text{lag}_0 \, \ln (\bar{E}/\bar{\pi}) + \bar{\beta}_8 K^4 \, \text{lag}_1 \, \ln (P^{eu}/\bar{\pi}) \\
&+ \bar{\beta}_9 K^2 \, \text{lag}_1 \, \ln (\bar{P}^c/\bar{\pi}) + \bar{\beta}_{10} K^3 \, G \, \text{lag}_1 \, \ln (\bar{P}^g/\bar{\pi}) + \bar{\beta}_{11} K^3 G \\
&+ \varepsilon
\end{aligned}
\tag{8.2.1}
$$

where the ownership of cookers, Z^3, is measured by the proportion of consumers owning one, A^3, and the temperature variable, W^u, is calculated using a threshold of $11 \cdot 7$ °C. Lag weights w_1 from a lognormal distribution with mode $1 \cdot 5$ are used throughout the following analysis for all the price variables. The terms K^k are weighted averages of the various ownership levels as defined by (6.6.2).[1] The other terms were described in 6.1.

It thus incorporates the results of Chapter 6 on the lag distribution, temperature threshold, and cooker ownership variable.

8.3 The electricity price variable P^{eu}

Given that the tariff function defines the marginal price in general as a decreasing step function of the quantity supplied in a quarter, it is possible that, if a household's actual demand deviates from the demand it was expecting to have, then the marginal rate may differ from the expected rate which determined its behaviour; and it is also possible that the actual rate would vary seasonally. Now for most households these possibilities would have been remote since their demand was always well above the level at which the final rate P^{ef} started to apply. For them, the current price which determined their behaviour, P^{eu}, may safely be taken to be the actual rate P^{ef}. But, for other consumers, to work out what this behaviour-determining price was some assumptions are necessary about how expectations of demand were formed and how consumers adjusted to the corresponding variations in the expected marginal price. The following assumptions are made:

(1) The weights used to calculate K^k are those from the 26-variable regression starred in Table 6.4.1, as these are the best prior estimates available. For computational reasons, it was impracticable to use the 26-variable equation here as it contains four terms in the price of electricity instead of one.

Assumption 8.3.1

Unexpected changes in any variable during a quarter did not influence expected price such as to affect demand later in the quarter.

This is simply an amplification of our earlier hypothesis that households will be slow to perceive and adjust to changed circumstances.

Hence only the formation of expectations at the beginning of a quarter need be considered.

Assumption 8.3.2

The expected values of prices other than of electricity, of total consumption expenditure and of ownership levels which enter the expressions for expected price and expected quantity demanded are equal to their actual values.

A lag in perceiving changes in these prices and in total expenditure has already been allowed for by the inclusion of lagged values and the giving of a relatively small weight to the current value. Whilst actual acquisitions of appliances by all consumers may have deviated from expected acquisitions, the discrepancy is unlikely to have been significant as a proportion of the total stock of all consumers.

Assumption 8.3.3

The habits which determined the rate of substitution of electricity for other fuels and other commodities in response to changes in natural conditions remained the same throughout any given year irrespective of whether the marginal price varied seasonally—even if this variation was expected. Furthermore, they were the same in all years even though the average temperature in particular years varied from the long-run average.

This implies that the price $(P^{eu})^h$ which appears in a consumer's demand function should be that rate which would be the marginal rate in "normal weather conditions" in a given Area. The long-run averages of temperature, $\overline{W}_B$, and hours of daylight, $\overline{L}_B$, are defined as the means for the period April 1952 to March 1968. The consequence of invalidity of this assumption would be that the coefficients of temperature and hours of daylight would be too high absolutely.

Assumption 8.3.4

The path taken to equilibrium following a change in the electricity tariff is the same as postulated previously when the final rate P^{ef} was used, that is, one which reaches equilibrium after twelve quarters *ceteris paribus*.

Assumption 8.3.5

After adjustment in accordance with the preceding assumptions, the mean of the expectations held by consumers at the beginning of a quarter with respect to the quantities they will demand during the quarter is equal to the mathematical expectation of the mean actual quantity demanded.

Let $(Y^p_{qB})^h$ be the expected quantity demanded per day of consumer h in quarter q which, given assumptions 8.3.1-8.3.4 and the tariffs ruling during that quarter in Area B, corresponds to the expected marginal price $(P^{eu})^h$ which appears in the expression for the consumer's actual demand.

Then assumptions 8.3.1-8.3.5 imply that the mean of the $(Y^p_{qB})^h$ over all consumers in Area B is equal to

$$
\begin{aligned}
Y^p_{qB} \equiv Y^u_{qB} &- \beta_{3,1} Z^1_{qB} \ln (W^u_{qB}/\overline{W}_B) \\
&- \beta_{3,2} Z^2_{qB} \ln (W^u_{qB}/\overline{W}_B) \\
&- \beta_{4,4} Z^{41}_{qB} \ln (L_{qB}/\overline{L}_B) \\
&- \bar{\beta}_8 K^4 [\text{lag}_1 \ln (P^{eu}/\bar{\pi})_{qB} - \ln (P^{eu}/\bar{\pi})_{qB}] \\
&- \varepsilon_{qB}
\end{aligned}
\tag{8.3.1}
$$

Two assumptions are made on the distribution about this mean:

Assumption 8.3.6

The distribution of values of expected demand $(Y^p_{qB})^h$ is uniquely defined by its mean.

Assumption 8.3.7

The distribution of values of actual demand per day is uniquely defined by its mean Y^u; in any given Area, the distribution of values of the expected demands $(Y^p_{qB})^h$ with mean equal to k, say, is identical to the distribution of values of actual demand which has mean $Y^u = k$.

As actual demand adjusted to normal weather conditions had a monotonic trend which was close to the further adjusted quantity variable Y^p, no substantial error should arise from using these two assumptions to determine the average expected marginal price from the observed relationship between the average marginal price actually paid and the actual mean quantity demanded.

8.4 The aggregate tariff function: data

The data on tariffs are complete, consisting for each tariff of a set of rates and block sizes with the variations, where applicable, in the primary rate as between prepayment and credit customers and in block sizes according to the number of rooms or floor area. Hence the set of rates and block limits which define the tariff function can be calculated for households with any given size of dwelling and type of meter.

The difficulty arises in the calculation of the aggregate relationship between the mean quantity supplied and the mean marginal rate—a relationship which will be called the *aggregate tariff function*. For this ideally data would be used both on the proportion of consumers in each Area in each quarter with any given tariff function and on the frequency distribution of the *unrestricted* consumption of each group with the same tariff function. In practice only two frequency distributions are available, of the *total* consumption per consumer in England and Wales as a whole in 1954/55 and 1965/66. Only the mean size of dwellings in each Area is available.

Let the mean marginal rate given the tariffs ruling at time t in Area B be P_{tB}^{em}. On the assumptions made above, this is a function only of the mean quantity supplied. Given the mean consumption in each Area in 1954/55 and 1965/66, P_{tB}^{em} for all t, B at these two consumption levels can be estimated from the data available on the following assumptions:

Assumption 8.4.1

No off-peak consumer had a total consumption such that the marginal unrestricted rate if all the units had been on unrestricted tariffs would have been different from his actual marginal rate.

This should be nearly true because the ownership of other appliances by consumers who had off-peak heaters was typically high enough to mean that their unrestricted consumption exceeded the lower limit of the final block.

Assumption 8.4.2

The proportion of consumers in any given Area with an annual consumption less than any given percentage of the Area mean consumption was the same as the proportion of consumers in England and Wales as a whole with an annual consumption less than the same percentage of the England and Wales mean.

This could not hold exactly in all Areas. It is only of importance at the bottom end of the range, in determining the proportion below the upper limit of the primary block. A check showed that what errors there are in individual Areas very nearly cancel out when the national average is calculated.

Given these two assumptions, we can calculate the proportion of consumers with mean quarterly consumption in each year between the upper limits of each block, where the limits are those applicable to consumers with the Area mean size of dwelling. These are not necessarily the correct weights to apply to the rates when calculating the mean marginal rate: some consumers whose *mean* quarterly consumption was within the limits of one block would have consumed a quantity outside this range in at least one quarter; as there was a positive correlation between consumption and size of dwelling and hence block sizes, some small consumers would have tended to have a lower marginal rate than if their dwelling was of average size and some large ones a higher rate. One further assumption is therefore made.

Assumption 8.4.3

The net effect on the estimated proportion of consumers having any given marginal rate of ignoring seasonal variation in consumption and variations in block sizes by size of dwelling was nil.

The higher primary rate for prepayment customers on some tariffs was not used as the requisite information on the proportion of consumers having it as their marginal rate was lacking—the proportion is only known to have been small.

8.5 The aggregate tariff function: form

The question remains of what values of the mean marginal rate correspond to values of the mean quantity supplied other than those which occurred in 1954/55 and 1965/66.

There are two considerations relevant to the choice of a specific hypothesis. Firstly, the distribution in 1954/55 was unimodal and positively skewed, that in 1965/66 of similar relative dispersion but less positively skewed and with a much higher mean. Hence, since only the size of the lower tail of the distribution is relevant to determining the proportion having the primary rate as their marginal rate, the aggregate tariff function can be expected to be non-linear—successive equal rises in the mean quantity supplied producing successively smaller falls in the mean marginal rate.

Secondly, writing the aggregate tariff function as

$$f(P_{tB}^{em}) = g(Y_{tB}^{u}) \qquad (8.5.1)$$

where f and g are functions to be defined, by the assumptions in 8.3, the same functions define the relationship between the mean expected marginal rate and the corresponding expected quantity demanded:

$$f(P_{tB}^{eu}) = g(Y_{tB}^{P}) \qquad (8.5.2)$$

Thus it is necessary to choose f and g such that a tractable reduced form is obtained when (8.5.2) is substituted into the demand equation.

The aggregate tariff function giving the mean actual marginal rate is therefore postulated to be

$$\ln P_{tB}^{em} = R_{tB} + S_{tB} Y_{tB}^u \qquad S_{tB} \leqslant 0 \tag{8.5.3}$$

where R_{tB} and S_{tB} are the same during all quarters t when the same set of tariffs were in operation. They are calculated by the method of simultaneous equations from the average values of Y_{tB}^u and P_{tB}^{em} in 1954/55 and 1965/66.

The form is obviously incorrect at the upper end of the possible range of Y^u since, for sufficiently high Y^u, it gives an average P^{em} less than the final (minimum) rate, P^{ef}. But at all the values of Y^u adjusted to normal weather conditions which occurred in the observation period, it gave an average greater than the final rate.

No error term is included in the equation since, if the assumptions of 8.4 are correct, the errors at the two datum points are zero and if the postulated functional form is correct, the error at any value of Y^u is zero. Whilst there may be non-zero errors due to invalidity of either of these premises—probably of a systematic kind—they should be smaller than errors in many of the other variables in the demand equation and the errors in that equation.

Thus the mean expected marginal price P^{eu} in the demand function is given by

$$\ln P_{tB}^{eu} = R_{tB} + S_{tB} Y_{tB}^p \tag{8.5.4}$$

where Y^p is the expectation of quantity demanded defined by (8.3.1). This will be called the *price function*.

8.6 The reduced form

The model comprises the simultaneous demand function (8.2.1) and price function (8.5.4).

The demand equation may be rewritten as

$$Y_{qB}^u = \sum_{j=1}^{15} \beta_j X_{qBj} + \varepsilon_{qB} \tag{8.6.1}$$

where
$$X_1 = 1$$
$$X_2 = Z^1$$
$$X_3 = Z^2$$
$$X_4 = Z^3$$
$$X_5 = Z^4$$
$$X_6 = Z^1 \ln W^u$$
$$X_7 = Z^2 \ln W^u$$
$$X_8 = Z^{41} \ln (L/\bar{L})$$

$$X_9 = Z^1 Z^{o1}$$
$$X_{10} = Z^2 Z^{o2}$$
$$X_{11} = K^4 \, \text{lag}_0 \, \ln (\bar{E}/\bar{\pi})$$
$$X_{12} = K^4 \, \text{lag}_1 \, \ln (P^{eu}/\bar{\pi})$$
$$X_{13} = K^2 \, \text{lag}_1 \, \ln (\bar{P}^c/\bar{\pi})$$
$$X_{14} = K^3 G \, \text{lag}_1 \ln (\bar{P}^g/\bar{\pi})$$
$$X_{15} = K^3 G$$

and the β coefficients are renumbered correspondingly. In this notation, the reduced form may be written more simply:

Let
$$
\begin{aligned}
X^o_{qBj} &\equiv Z^1_{qB} \ln \overline{W}_B & j&=6 \\
&\equiv Z^2_{qB} \ln \overline{W}_B & j&=7 \\
&\equiv Z^{41}_{qB} \ln (\overline{L}_B/\overline{L}) & j&=8 \\
&\equiv X_{qBj} & &\text{otherwise}
\end{aligned} \right\}
\tag{8.6.2}
$$

and

$$
\Theta_{tB} \equiv \frac{1}{1 - S_{tB}\beta_{12}K^4_{tB}}
\tag{8.6.3}
$$

Then the reduced form is

$$
Y^u_{qB} = \sum_{j \neq 12} \beta_j \left(X_{qBj} + \beta_{12} K^4_{qB} \sum_{t=q-11}^{q} w^1_{t-q} \Theta_{tB} S_{tB} X^o_{tBj} \right)
$$

$$
+ \beta_{12} K^4_{qB} \sum_{=q-11}^{4} w^1_{t-q} \widehat{\Theta}_{tB} (R_{tB} - \ln \bar{\pi}_{tB})
$$

$$
+ \varepsilon_{qB}
\tag{8.6.4}
$$

8.7. Estimation and testing: methods

Ex hypothesi the error ε_{qB} has the same properties as in the earlier model and the variance-covariance matrix is $\sigma^2 V$ as defined in 5.3. Therefore, writing the estimated reduced form as

$$
Y^u_{qB} = \sum_{j \neq 12} b_j \left(X_{qBj} + \hat{\beta}_{12} K^4_{qB} \sum_t w^1_{t-q} \widehat{\Theta}_{tB} S_{tB} X^o_{tBj} \right)
$$

$$
+ b_{12} K^4_{qB} \sum_t w^1_{t-q} \widehat{\Theta}_{tB} (R_{tB} - \ln \bar{\pi}_{tB})
$$

$$
+ e_{qB}
\tag{8.7.1}
$$

where
$$
\widehat{\Theta}_{tB} = \frac{1}{1 - S_{tB}\hat{\beta}_{12}K^4_{tB}}.
\tag{8.7.2}
$$

b_j ($j=1, 2, \ldots, 15$) and $\hat{\beta}_{12}$ were chosen to minimize $e'V^{-1}e$ subject to the constraint $\hat{\beta}_{12}=b_{12}$. All 624 observations were used.

Descriptive and test statistics have been calculated as in Chapter 5, as if $\hat{\beta}_{12}$ were known with certainty *a priori* and not necessarily equal to b_{12}. This should not seriously affect comparisons with earlier results.

8.8 Results

The full results of estimating the reduced form are given in Table 8.8.1. To facilitate comparison, the coefficients and F-values are given again in Table 8.8.2 alongside those of the earlier single-equation model using the final rate P^{ef}.

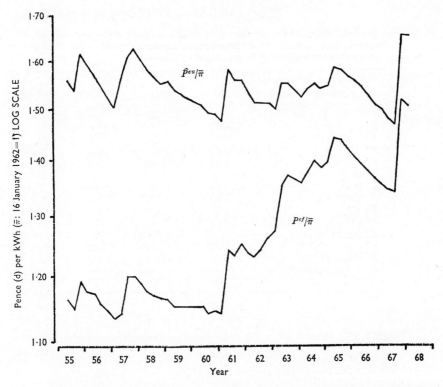

FIGURE 8.8.1. *The relative price of electricity in England and Wales*

$\hat{P}^{eu}$: an average of all the unrestricted rates, as defined in 8.8.

P^{ef} : the final rate on the standard domestic unrestricted tariffs.

The regression estimate $\hat{P}^{eu}$ of the average expected marginal price is compared with the final rate P^{ef} in Figure 8.8.1. The rise in the estimated average in England and Wales was so much less than in the final rate that relative to other prices the estimated average shows virtually no upward trend whereas the final rate rose quite rapidly. Several factors contribute to the explanation of this:

(1) there was a large rise in mean consumption;

(2) the distribution of consumption became less positively skewed, making for an even greater decline in the proportion with low consumption;

(3) a small decline in the proportion with consumption below the upper limit of the primary block had a large effect on the average marginal

TABLE 8.8.1. *Tariff function model: results*

Brief definitions of the variables appearing were given in 6.1, and full definitions in Chapter 4. K^2, K^3, K^4 are weighted averages, defined by (6.6.2), of the ownership levels of the different classes of appliances.

Variable k in the demand equation (8.6.1)	b_k	$\hat{\sigma}_k$	$F\{b_k\}$	$F\{b_k^o\}$
1	0·434	0·352	1·52	90302
Z^1	32·2	2·34	189·3	6217
Z^2	−25·3	4·88	26·93	1291
Z^3	−16·7	2·39	48·77	95·34
Z^4	−1·27	0·886	2·06	272·3
$Z^1 \ln W^u$	−5·99	0·370	261·8	6336
$Z^2 \ln W^u$	1·66	0·738	5·06	2·98
$Z^{41} \ln (L/\bar{L})$	−1·50	0·242	38·51	32·80
$Z^1 Z^{o1}$	−3·13	0·266	138·8	171·2
$Z^2 Z^{o2}$	5·55	2·31	5·75	17·38
$K^4 \lag_0 \ln (\bar{E}/\bar{\pi})$	0·245	0·0981	6·23	15·38
$K^4 \lag_1 \ln (P^{eu}/\bar{\pi})$	0·3295	0·0326	102·0	131·2
$K^2 \lag_1 \ln (\bar{P}^c/\bar{\pi})$	2·78	0·306	82·47	121·5
$K^3 G \lag_1 \ln (\bar{P}^g/\bar{\pi})$	−0·00811	0·140	0·00	15·31
$K^3 G$	1·77	0·504	12·33	12·33

$$R^2 = 0.958 \qquad\qquad F\{R^2\} = 1052$$

TABLE 8.8.2. *Results when the price function is omitted: comparison with the other results*

The F-values are given in brackets after the coefficients

Electricity price variable P^e	Final rate P^{ef}		Average P^{eu} Price function included		Average P^{ev} Price function omitted	
Equation	(6.6.1)		(8.6.4)		(8.9.1)	
1	0·616	(4·57)	0·434	(1·52)	1·34	(20·67)
Z^1	31·4	(192·7)	32·2	(189·3)	31·2	(189·2))
Z^2	−32·0	(47·08)	−25·3	(26·93)	−33·1	(49·22)
Z^3	−17·7	(73·08)	−16·7	(48·77)	−15·8	(61·13)
Z^4	−3·00	(14·91)	−1·27	(2·06)	−2·45	(10·60)
$Z^1 \ln W^u$	−6·06	(283·3)	−5·99	(261·8)	−6·11	(286·4)
$Z^2 \ln W^u$	1·75	(5·92)	1·66	(5·06)	1·83	(6·44)
$Z^{41} \ln (L/\bar{L})$	−1·51	(41·50)	−1·50	(38·51)	−1·49	(39·72)
$Z^1 Z^{o1}$	−3·27	(248·6)	−3·13	(138·8)	−3·42	(265·8)
$Z^3 Z^{o2}$	6·61	(12·18)	5·55	(5·75)	6·95	(13·29
$K^4 \lag_0 \ln (\bar{E}/\bar{\pi})$	0·481	(30·77)	0·245	(6·23)	0·421	(25·31)
$K^4 \lag_1 \ln (P^e/\bar{\pi})$	−0·203	(18·86)	0·3295	(102·0)	−0·157	(13·56)
$K^2 \lag_1 \ln (\bar{P}^c/\bar{\pi})$	3·66	(216·8)	2·78	(82·47)	3·91	(215·6)
$K^3 G \lag_1 \ln (\bar{P}^g/\bar{\pi})$	−0·171	(2·54)	−0·00811	(0·00)	−0·313	(7·82)
$K^3 G$	1·24	(8·77)	1·77	(12·33)	1·45	(12·08)
R^2	0·960		0·958		0·960	

with the reduced form equation (8.6.4)

Dummy variable k	c_k	$F\{c_k\}$	$F\{c_k^o\}$	d_B
1	1·46	3·64	0·02	
Moving-quarter:				
II	0·0544	0·01	1·19	
III	−0·375	0·45	0·05	
IV	−0·956	2·92	2·92	
Area:				
South Western				1·07
London	1·82	3·52	19·95	1·83
South Eastern	−2·78	8·21	4·57	0·51
Southern	−1·54	2·52	0·21	1·66
Eastern	−2·52	6·76	4·13	1·69
East Midlands	−2·29	5·56	3·76	1·58
Midlands	−0·28	0·09	0·84	1·33
South Wales	−2·00	4·27	2·76	1·48
Merseyside and North Wales	−0·11	0·01	1·09	1·61
Yorkshire	0·60	0·38	6·49	0·76
North Eastern	−0·46	0·23	2·90	0·84
North Western	—3·79	15·29	15·29	1·49

$$F\{c_2^o, c_3^o, c_4^o\} = 1·38$$
$$F\{c_5^o, \ldots, c_{15}^o\} = 5·64$$

Autocorrelation coefficients: $\hat{\rho} = 0·345$ $t\{\hat{\rho}\} = 8·96$

$\hat{\rho}_q = 0·687$ $t\{\hat{\rho}_q\} = 10·27$

Heteroscedasticity coefficient: $\hat{\psi} = 0·366$ $t\{\hat{\psi}\} = 3·05$

TABLE 8.8.2 *continued*

Zero-order correlation coefficients:

	p^{ef}	$\hat{p}^{eu}$	p^{ev}
p^{ef}	1·000	—	—
$\hat{p}^{eu}$	·697	1·000	—
p^{ev}	·678	·970	1·000

rate since, at the beginning of the period, the primary rate was five times the final rate on average;

(4) the size of primary blocks did not rise in proportion to mean consumption and the primary rate rose less than the final rate on average, being only about four times the final rate by the end of the period.

The most notable change in the regression results compared with those with the single-equation model is that the coefficient of the price of electricity is now significantly *positive* with a higher F-value, and the coefficients of the price of solid fuel and total expenditure are lower with lower F-values. The reason for this is much clearer when we look at the results with a third model in which, incorrectly, an average of the rates is used but no price function included.

8.9 Omission of the price function

The equation estimated, by the direct application of Generalized Least Squares, was

$$Y^u = \sum_{j \neq 12} \alpha_j X_j + \alpha_{12} K^4 \text{ lag}_1 \ln (P^{ev}/)\bar{\pi} + \varepsilon \qquad (8.9.1)$$

where the X_j are defined as in 8.6 and P^{ev} is the estimated average marginal price of electricity in normal weather conditions given by

$$\ln P^{ev}_{qB} = R_{qB} + S_{qB}[Y^u_{qB} - b_6 Z^1_{qB} \ln (W^u_{qB}/\overline{W}_B)$$
$$- b_7 Z^2_{qB} \ln (W^u_{qB}/\overline{W}_B)$$
$$- b_8 Z^{41}_{qB} \ln (L_{qB}/\overline{L}_B)] \qquad (8.9.2)$$

The coefficients b_6, b_7, and b_8 are the estimates obtained using the constrained demand equation (6.6.1).

P^{ev} is defined in this way so that it could be calculated prior to estimation of (8.9.1). As a result, it differs from the mean expected marginal rate P^{eu} defined by (8.5.4) in having no terms in the price of electricity or the disturbance on the right-hand side. $\hat{P}^{eu}$, the regression estimate of P^{eu} is, however, very highly correlated with P^{ev} so that this discrepancy should not affect comparisons appreciably. The correction to normal weather is made to keep the definitions as close as possible; as was borne out by the results, the estimated coefficients of temperature and hours of daylight, b_6, b_7, and b_8, from the earlier single-equation model could be taken to be very close to the estimates of the coefficients of the same variables obtained from (8.9.1).

The results are shown in the final column of Table 8.8.2. Comparison with the results with the simultaneous equations model shows that the consequence of omitting the price function is that there is such a large downward bias in the coefficient of the price of electricity that it becomes significantly negative; the coefficients of the price of solid fuel and total expenditure are correspondingly upward biased; the apparent explanatory power of the term in the price of electricity is reduced and that of the terms in the price of solid fuel and total expenditure increased.

A striking feature of the results with this incorrectly specified model is their similarity to the results with the single-equation model using the final rate P^{ef}. Some conclusions can be drawn from this about the explanation of the differences between the results when using the final rate P^{ef} and those when using the average P^{eu}.

8.10 Conclusions

Comparing the three sets of results, we conclude that, although the final rate P^{ef} is not itself dependent on the quantity supplied—at least in the short run—it is so highly correlated with the correct average of the rates, which is, that the effect of the combination of the correlation and the omission of the price function is simultaneous equation bias.

This probably does not explain all the difference in the results between the single equation model using the final rate P^{ef}, and the simultaneous equation model using the average marginal rate, P^{eu}: firstly, the analysis of Chapter 3 indicated that a mean of the rates weighted by the ownership levels should be used; in the average calculated the rates are weighted instead by the number of consumers. Given the positive correlation between ownership and demand, the correct mean, on the assumptions of Chapter 3, would lie between the final rate and the calculated average and not necessarily closer to this average than to the final rate. If, contrary to assumption 3.4.1 on the distribution of the coefficients $\beta^h_{j,a}$, those consumers whose marginal rate was the final rate were more sensitive to price changes then the correct mean would tend to be closer to the final rate than to the calculated average. This would be so if say the richer households tended to own more substitute appliances and therefore have larger U effects, and if the ownership of poorly-utilized appliances in any class was more elastic with respect to the fuel price amongst richer households so that Z-U effects were larger.

Secondly the additional data used in the calculation of the average is less reliable than the basic tariff data and so errors are likely in the average whereas there should be none in the series for the final rate.

Nevertheless, what bias there is in the simultaneous equations model due to these two causes should be much smaller than the simultaneous equation bias in the single-equation model. We therefore conclude that the regression results in Table 8.8.1 for the tariff function model are an improvement on those in Table 6.6.1 for the single equation model using the final rate.

Most of the derived results of 6.7-6.13 still stand, however: the goodness of the explanation of growth rates, inter-Area differences, and the seasonal pattern is virtually as described in 6.7; the pattern of the residuals has the same general features so that the conclusions of 6.8 on likely specification errors and the importance of errors in the variables still hold. Within the margins of error indicated in 6.11, the estimates of the trend and seasonal pattern in the utilization of each class of appliance are the same. Demand in the South of Scotland is still underpredicted. The conclusion reached in 6.13 that a rise in the price of electricity had a negative Z effect is strengthened.

The better estimates of the β-elasticities from the simultaneous equations model are given in Table 8.10.1. Compared with the earlier estimates given in Table 6.9.1, the elasticities with respect to the total expenditure ($\bar{E}$), the price of solid fuel ($\bar{P}^c$), the price of gas ($\bar{P}^g$) and gas availability (G) show large proportional changes in their numerical values; but the signs of the implied Z-U effects are the same as deduced in 6.10 and hence so are the interpretations to be placed on the coefficients.

TABLE 8.10.1. *Unrestricted demand: β-elasticities estimated from equation (8.6.1)*[1]

Variable	β-elasticity	Variable	β-elasticity
$\bar{E}$	0·23	W^u	1·19
P^{eu}	0·31	L	−0·12
$\bar{P}^c$	0·55	Z^{o1}	−0·040
$\bar{P}^g$	−0·0022	Z^{o2}	−0·0032
G	0·48		

(1) Calculated at the sample means defined in 6.9 and
therefore comparable with Table 6.9.1.

There remains to be interpreted the coefficient of the price of electricity. In the terminology of 6.1, this coefficient and the elasticity given in Table 8.10.1, like those discussed earlier, measure the sum of the U effect and the deviation of the Z-U effect from the average for all changes in ownership whatever the cause. It is therefore quite possible for the total effect, Z effects and U effects to be of opposite sign to the coefficient. Thus the positive sign of the coefficient is consistent with negative elasticities of demand for un-restricted electricity, of demand for all appliances and of utilization of all appliances given the stock of appliances. The plausible explanation of the sign and magnitude of the coefficient is that a fall in the relative marginal price of electricity had a negative and fairly large Z-U effect: low utilization households acquired appliances in some classes for the first time, or existing owners added to their stock of appliances such that at least some of the augmented stock was used less intensively than the original stock had been.

CHAPTER 9

CONCLUSION

> "Finally, we must never forget that our progress in understanding economic laws depends strictly on the quality and abundance of statistical data. Nothing can take the place of the painstaking work of objective observation of the facts."
>
> E. MALINVAUD (1966): *Statistical Methods of Econometrics*

9.1 Characteristics of the analysis

Our aim has been to explain the variations in domestic electricity consumption over time and between areas. The starting-point was a simple model based on the ordinary classical demand theory.[1] It was argued that modifications to this model were desirable to take account of restrictions deduced from the theoretical postulates and known characteristics of the electricity market.[2] After examining the limitations imposed by the data[3] and the models used in previous studies,[4] it was concluded[5] that a feasible and promising approach was:

(1) to concentrate on analysing the effects, both long- and short-run, on mean utilization of changes in prices, income, and natural conditions, taking appliance ownership as given and predetermined;

(2) to incorporate six of the modifications:
   (i) to include both hours of daylight and temperature to represent natural conditions, and to allow for non-linearity of response;
   (ii) to disaggregate by type of appliance;
   (iii) to allow, in the form of the equation, for the complementarity of the fuel and the appliances;
   (iv) to allow for a lagged response to changes in prices and income;
   (v) to disaggregate into unrestricted and off-peak demand;
   (vi) in the variant model in Chapter 8, to allow for the dependence

(1) See 1.6.
(2) See 1.7.
(3) See 1.8.
(4) See 2.2-2.11.
(5) See 2.12.

of the average marginal price of electricity on the quantity supplied in the short run.

As it was believed that the results of the analysis would be sensitive to the details of the definitions of the variables used, particular attention was paid to getting these definitions as close as the data permitted to those derived from the postulates of the theory and empirical evidence.[6]

9.2 The results achieved

The results of the analysis have shown that there *are* variations in the household demand for electricity which are not simply related to changes in the ownership of appliances or natural conditions, and that ordinary demand theory, suitably modified, *is* capable of explaining at least a large part of the implied variations in mean utilization.[7] As seen from Tables 6.7.1 and 7.20.1, the model explains fairly well the absolute level of consumption per consumer in each Area, the rate of growth of off-peak sales and the downturn in the rate of growth of consumption on unrestricted tariffs in the latter part of the period. The explanation of growth rates of unrestricted consumption over short periods in individual Areas is poor but this could be due to errors in the ownership variables, as the movements in the ownership series over such short periods are open to considerable doubt.[8]

Including hours of daylight after 06.30 hours as a measure of illumination and postulating that sensitivity to temperature changes declined, as temperature rose, to zero above an estimated threshold were both found to improve significantly the explanatory power of the model. Off-peak demand was found to be sensitive to temperature changes up to a higher threshold than unrestricted demand.[9]

Disaggregation by type of appliance posed problems in obtaining sufficiently reliable estimates of parameters,[10] because of the correlation between the ownership series, but was worthwhile: the variations in the derived estimates of the mean utilization of each class of appliances show that the assumptions behind an aggregate model were far from valid in the period 1955-68.[11]

(6) See 4.1 and the rest of Chapter 4. In the course of the analysis, this sensitivity was confirmed in the case of the cooker ownership and electricity price variables—see 6.4 and 8.10 respectively.

(7) Compare 1.1.The most acceptable estimates of individual parameters and measures of their reliability are given in Table 8.8.1 for unrestricted demand (Table 7.19.1 for off-peak demand) and are interpreted in 6.9, 6.10 and 8.10 (7.22 for off-peak demand).

(8) See 4.11.

(9) See 6.3 and 7.16.

(10) See 6.2.

(11) See 6.11.

The form of the complementarity between the fuel and the appliances means that the effect on demand of a change in the stock of appliances depends on the values of all the arguments of their utilization functions and that the effect of a change in one of these arguments depends on the size of the stock. Given the range of the values of the variables observed, it would therefore have been a serious error to specify the demand function as linear in all its arguments. To produce a tractable regression equation, with dis-aggregation of the stock of appliances, the complementarity had to be allowed for in a way which led to the appearance of the same variable in several terms in the equation, and hence to multicollinearity. But this was less serious since constraints could be introduced[12] and many hypotheses could be tested by joint F-tests of the coefficients of intra-correlated groups of terms.

From trying three alternative lag distributions, the tentative conclusion can be reached that most of the adjustment of utilization to changes in prices took place fairly quickly. The best results were achieved with a distribution which implies a maximum rate of adjustment in the quarter following that in which the price changed and virtually complete adjustment within two years.[13]

The division into unrestricted and off-peak demand revealed that the two had quite different characteristics: variations in off-peak demand were almost entirely accounted for by variations in appliance ownership, utilization remaining constant, and in natural conditions,[14] whereas there were variations in the mean utilization of unrestricted appliances of which a significant part is explained by variations in prices, income, and the availability of gas.[15] An explanation of the virtual absence of such variations in off-peak utilization has been advanced in 7.22.

There are two important conclusions from the analysis in Chapter 8 about the use of an average marginal price of electricity:[16]

(1) the results with the variant model, in which an average is used and its dependence on the quantity supplied through the tariff structure is allowed for by including a function describing the tariff, are an improvement over those with our earlier model in which the final rate on the standard unrestricted tariffs was used;

(2) using an average without including a second equation to describe its dependence on the quantity supplied through the tariff structure leads to seriously biased coefficients.

(12) See 6.2 and 6.5.
(13) See 6.4 and 7.17.
(14) See 7.20.
(15) See 6.10.
(16) See 8.10.

9.3 Future research

The integration of the utilization functions into a model including appliance demand functions thus promises to be a fruitful subject for future research. For its estimation, more data—on taste variables and the prices of appliances—will need to be assembled.

Further improvement of the estimates of the utilization parameters must depend mainly on improvements in "the quality and abundance of statistical data". If the model could be estimated at a lower level of aggregation—over both consumers and types of appliances—the assumptions could be relaxed and effects on the utilization of a given stock of appliances distinguished from those on mean utilization associated with changes in the stock. To make this possible, the greatest needs are for more information to be collected on the ownership of appliances and on the consumption of individual types of appliances; and for more data on the determinants of utilization to be collected from the same samples.

It is to be hoped that eventually more data will become available on other fuels and the appliances which use them so that the equations for electricity and electrical appliances can be estimated as part of a larger system.

APPENDIX

LIST OF NOTATION

The meanings attached to letters are briefly described here. If the same definition is used in all chapters, a reference to where the full definition is introduced is given in the final column. A reference followed by "& Ch. 2" means that the same definition applies in all chapters except Chapter 2 where similar but not identical definitions are used.

Letters marked "local" are used to denote terms which only appear in one chapter or section. They may subsequently be redefined. Subscripts, superscripts, and symbols attached to a letter are only given here if they alter its general meaning.

Distinct meanings are attached to bold letters, which denote matrices.

Letter	Meaning	Reference
A	Proportion of consumers having appliances in a specified class	3.2 & Ch. 2
a	Class of appliance	—
B	Electricity Board Area	—
b, $\mathbf{b}$	Estimator of coefficient β ($\boldsymbol{\beta}$) in demand equation	5.3 & Ch. 2
b^0, $\mathbf{b}^0$	Estimator of coefficient(s) in orthonormalized demand equation	5.4
C	Number of domestic electricity consumers	4.2 & Ch. 2
c, $\mathbf{c}$	Estimator of coefficient γ ($\boldsymbol{\gamma}$) in regression of residuals on seasonal and Area dummies	5.6
c^0, $\mathbf{c}^0$	Estimator of coefficient(s) in the orthonormalized regression of residuals on seasonal and Area dummies	5.6
D^m	Number of days in month m	4.3
$\mathbf{D}$	Observations matrix in regression of residuals on seasonal and Area dummies	5.6
d		local
d_B	Durbin-Watson d statistic for Area B time-series	5.7

∂	Partial differential operator	—
E	Followed by {...}: mathematical expectation	—
	Otherwise: total consumption expenditure per household	4.17 & Ch. 2
$\bar{E}$	Total consumption expenditure per household, seasonally adjusted	4.19
e, **e**	Residual(s) in estimated demand equation	5.4
e$^\circ$	Residuals in estimate of orthonormalized demand equation	5.4
$F(...)$	F-distribution (degrees of freedom in brackets)	—
$F\{...\}$	F-test statistic (estimators of the coefficients hypothesized to be zero in brackets)	5.5
F	Lower triangular matrix: $\mathbf{D'.D} = \mathbf{F.F'}$	5.6
f		local
G	Proportion of electricity consumers having a gas supply available	4.12
g		local
H		local
H_0	Hypothesis under test	—
H_1	Alternative hypothesis	—
h	As superscript: consumer	—
	Otherwise:	local
I	Income per household	4.17 & Ch. 2
I	Identity matrix	—
i		local
J		local
j		local
K^0	Weighted sum of Z^{o1} and Z^{o2}	7.19
K^2	Weighted sum of Z^1 and Z^2	6.6
K^3	Weighted sum of Z^1, Z^2 and Z^3	6.6
K^4	Weighted sum of Z^1, Z^2, Z^3, and Z^4	6.6
k		local
L	Minutes of daylight after 06.30 hours	4.20 & Ch. 2
$\bar{L}$	Mean value of L	6.2
l	Mean wattage rating of appliances in specified class per consumer having	3.2
lag	Operator: weighted average of current and lagged values of operand	3.5
ln	Operator: natural logarithm	—
log	Operator: logarithm	—
M_q^m	Proportion of month m's consumption read in quarter q	4.3

m	As subscript or superscript: month	—
	Otherwise: number of observations	—
N		local
n	Number of variables	—
P^c	Price of solid fuel	4.16 & Ch. 2
$\bar{P}^c$	Price of solid fuel, seasonally adjusted	4.19
P^e	Price of electricity	4.14 & Ch. 2
P^{ef}	Final rate on standard domestic unrestricted electricity tariffs	4.14 & Ch. 2
P^{em}	Average marginal price of electricity actually experienced	8.4
P^{eo}	Price of off-peak electricity	7.12
P^{eu}	Average marginal price of electricity expected by consumers in normal weather conditions	8.3
P^{ev}	Regression estimate of average marginal price of electricity in normal weather conditions	8.9
P^g	Price of gas	4.15 & Ch. 2
$\bar{P}^g$	Price of gas, seasonally adjusted	4.15
$\mathbf{P}$	Lower triangular matrix: $\mathbf{V} = \mathbf{P}.\mathbf{P}'$	5.4
P	Price	local
Q	Total quantity of electricity consumed	4.2 & Ch. 2
Q^o	Quantity of electricity purchased on off-peak tariffs	4.2
Q^u	Quantity of electricity purchased on unrestricted tariffs	4.2
q	As subscript: quarter	—
	Otherwise: quantity	local
R	Intercept in tariff function	8.5
R^2	Coefficient of determination	5.5
$\mathbf{R}$	Inverse of $\mathbf{P}$	5.4
r	Followed by (...): zero-order correlation coefficient	5.5
	Otherwise: mean proportion of the wattage rating of appliances in a specified class in use when the services of appliances are demanded or are on to meet a later demand	3.2
S	Slope of tariff function	8.5
$\mathbf{S}$	Lower triangular matrix: $\mathbf{X}'.\mathbf{X} = \mathbf{S}.\mathbf{S}'$	5.4
s	Element of $\mathbf{S}$	5.4
$\hat{T}$	Regression estimate of the identifiable part of a utilization function	6.4
$\bar{T}$	Sample mean value of $\hat{T}$	6.4

T	Inverse of **S**	5.4
t	With subscripts: element of **T**	5.4
	Followed by {...}: t-test statistic (estimator of the coefficient hypothesized to be zero in brackets)	5.5
	Otherwise: time	local
U	*Either* demand in kWh due to use of appliances in a specified class, per consumer having; *or* demand in hours per kW of installed load of appliances in a specified class—according to the definition of the corresponding ownership variable *Z*	3.3 & Ch. 2
u	Time, in hours, during which the services of appliances in a specified class are demanded or the appliances are on to meet a later demand	3.2
V	Expenditure on a specified fuel at constant prices	local
V	Error variance-covariance matrix	5.3
v	Element of **V**	4.5
W	Temperature	4.20 & Ch. 2
W^o	The lower of actual temperature and the estimated threshold $\hat{\tau}^o$	7.13
W^u	The lower of actual temperature and the estimated threshold $\hat{\tau}^u$	4.20
w	Lag weight	4.19
X	Explanatory variable	3.5
X	Observations matrix	5.2
x	Element of **X**	5.2
x	Column of **X** (observations on one variable)	5.2
Y^o	Mean quantity demanded of electricity on off-peak tariffs per consumer per day	7.8
Y^p	Mean quantity of electricity on unrestricted tariffs per consumer per day which would be demanded if the price of electricity were as expected by consumers in normal weather conditions	8.3
Y^u	Mean quantity demanded of electricity on unrestricted tariffs per consumer per day	4.4
Y	Dependent variable	5.2
y		local
Z	Ownership of appliances in a specified class—*either* the proportion of consumers having, *or* installed load per consumer in kW	3.3 & Ch. 2

$z, \mathbf{z}$	Residual(s) in the regression of the demand equation residuals on seasonal and Area dummies	5.6
α	Coefficient	local
$\beta, \boldsymbol{\beta}$	Coefficient(s) in demand equation	3.3, 5.2 & Ch. 2
Γ		local
$\gamma, \boldsymbol{\gamma}$	Coefficient(s) in the equation relating the demand equation errors to seasonal and Area dummies	5.6
Δ	Operator: forward first differences	—
Δ'	Operator: backward first differences	—
δ	Random error	5.8
$\varepsilon, \boldsymbol{\varepsilon}$	Error(s) in demand equation	3.3 & Ch. 2
$\zeta, \boldsymbol{\zeta}$	Random error(s)	5.6
η	Random error	local
Θ	Variable	local
θ	Parameter	local
κ	Parameter	local
Λ	Cumulative lognormal distribution	—
λ_q^m	Weight given to month m value in calculating the moving-quarter mean for quarter q	4.4
μ	Parameter	local
ξ		local
π	Index of the prices of all commodities except fuels	4.18 & Ch. 2
$\bar{\pi}$	Index of the prices of all commodities except fuels, seasonally adjusted	4.19
ρ	Autocorrelation coefficient	5.7
$\hat{\rho}$	Estimator of ρ	5.7
$\sum$	Operator: summation	—
σ	Standard deviation of error in demand equation	3.10
$\hat{\sigma}$	Estimator of σ	5.5
$\hat{\sigma}_k$	Estimated standard error of estimator b_k	5.5
$\hat{\sigma}_\xi$	Estimated standard error of ζ	5.6
$\hat{\sigma}_\rho$	Estimated standard error of $\hat{\rho}$	5.7
τ^o	Temperature above which demand for electricity on off-peak tariffs is temperature-insensitive	7.13
$\hat{\tau}^o$	Estimate of τ^o	7.16
τ^u	Temperature above which demand for electricity on unrestricted tariffs is temperature-insensitive	4.20
$\hat{\tau}^u$	Estimate of τ^u	6.3

υ	Random error	local
ϕ	Parameter	local
χ^2	χ^2-distribution	—
ψ	Heteroscedasticity coefficient	5.8
$\hat{\psi}$	Estimator of ψ	5.8
ω	Standard deviation	local

Numbering of appliance classes

The classes of appliances, as defined in 4.8-4.10, are numbered throughout as follows:

    1 Space-heaters on unrestricted tariffs
    2 Water-heaters on unrestricted tariffs
    3 Cookers
    4 Lighting and sundry appliances
       41 Lighting appliances
       42 Sundry appliances
    o1 Space-heaters on off-peak tariffs
    o2 Water-heaters on off-peak tariffs

Numbering of coefficients

The coefficients $\beta_{j,a}$ and estimators $b_{j,a}$ in the demand equation are numbered as below from Chapter 3 onwards. The same numbering is used in Chapter 2 where similar variables appear.

    j Variable some function of
    1 Reciprocal of ownership level (constant outside utilization function)
    2 Constant
    3 Temperature
    4 Hours of daylight
    5 Ownership of off-peak space-heaters
    6 Ownership of off-peak water-heaters
    7 Total consumption expenditure per household
    8 Price of electricity
    9 Price of solid fuel
    10 Price of gas
    11 Availability of a gas supply

LIST OF WORKS CITED

Statistical sources are listed separately on p. 152

AITCHISON, J. & J. A. C. BROWN (1957). *The Lognormal Distribution*. Cambridge University Press.

BAIN, A. D. (1964). *The Growth of Television Ownership in the United Kingdom*. Cambridge University Press.

BALESTRA, P. (1967). *The Demand for Natural Gas in the United States*. North-Holland, Amsterdam.

BAXTER, R. E. & R. REES (1968). Analysis of the industrial demand for electricity. *Economic Journal*, vol. 78.

BROWN, J. A. C. (1958). On the use of covariance techniques in demand analysis. Unpublished paper, University of Cambridge Department of Applied Economics.

BROWN, A. & A. DEATON (1972). Models of consumer behaviour: a survey. *Economic Journal*, vol. 82.

CRAMER, J. S. (1959). Private motoring and the demand for petrol. *Journal of the Royal Statistical Society*, Series A, vol. 122.

—— (1969). *Empirical Econometrics*. North-Holland, Amsterdam.

DAVIES, M. (1958). The relationship between weather and electricity demand. Institute of Electrical Engineers Monograph 314S, in *Institute of Electrical Engineers Proceedings*, Part C.

DOUMENIS, G. J. (1965). The demand for electricity in Great Britain: a study in econometrics. Unpublished Ph.D. thesis, University of Southampton.

DURBIN, J. & G. S. WATSON (1950). Testing for serial correlation in least squares regression. I. *Biometrika*, vol. 37.

FARRELL, M. J. (1954). Some aggregation problems in demand analysis. *Review of Economic Studies*, vol. 21.

FARRELL, M. J. (1959). The new theories of the consumption function. *Economic Journal*, vol. 69.

FISHER, F. M. & C. KAYSEN (1962). *A Study in Econometrics: the Demand for Electricity in the United States*. North-Holland, Amsterdam.

FORSTER, C. I. R. & I. J. WHITTING (1968). An integrated mathematical model of the fuel economy. Central Statistical Office: *Statistical News*, no. 3, November 1968.

FRIEDMAN, M. (1957). *A Theory of the Consumption Function*. Princeton University Press for National Bureau of Economic Research.

HOUTHAKKER, H. S. (1951). Some calculations on electricity consumption in Great Britain. *Journal of the Royal Statistical Society*, Series A, vol. 114.

—— (1952). The econometrics of family budgets. *Journal of the Royal Statistical Society*, Series A, vol. 115.

—— & L. D. TAYLOR (1966). *Consumer demand in the United States 1929-1970: Analyses and Predictions* (1st edition). Harvard University Press, Cambridge, Mass.

HURWICZ, L. (1950). Least squares bias in time-series. Chapter 15, pp. 365-383, in T. C.

KOOPMANS (ed.) *Statistical Inference in Dynamic Economic Models*. Cowles Commission Monograph 10, Wiley, New York.

IRONMONGER, D. S. (1972). *New commodities and consumer behaviour*. Cambridge University Press.

JOHNSTON, J. (1963). *Econometric Methods* (1st edition). McGraw-Hill, New York.

KLEITMAN, N. (1963). *Sleep and Wakefulness* (revised edition). University of Chicago Press.

LONGLEY. J. W. (1967). An appraisal of least squares programs for the electronic computer from the point of view of the user. *Journal of the American Statistical Association*, vol. 62.

LYNESS, F. K. & E. H. M. BADGER (1970). A measure of winter severity. *Journal of the Royal Statistical Society*, Series C, vol. 19.

MALINVAUD, E. (1966). *Statistical Methods of Econometrics* (1st edition). North-Holland, Amsterdam.

MEEK, R. L. (1968). The new Bulk Supply Tariff for electricity. *Economic Journal*, vol. 78.

NEEDLEMAN, L. (1960). The demand for domestic appliances. *National Institute Economic Review*, November 1960.

PEARCE, I. F. (1964). *A Contribution to Demand Analysis*. Oxford University Press.

PLACKETT, R. L. (1960). *Principles of Regression Analysis*. Oxford University Press.

PRAIS, S. J. & H. S. HOUTHAKKER (1955). *The Analysis of Family Budgets*. Cambridge University Press.

SAMUELSON, P. A. (1947). *Foundations of Economic Analysis*. Harvard University Press, Cambridge, Mass.

SHELL INTERNATIONAL PETROLEUM CO. (1969). Energy demand and supply in the U.K. 1975-80. *Economic Studies*, vol. 4.

STONE, R. (1951). *The Role of Measurement in Economics*. Cambridge University Press.

—— (1954). *The Measurement of Consumers' Expenditure and Behaviour in the United Kingdom 1920-1938*, Volume 1. Cambridge University Press.

—— & G. CROFT-MURRAY (1959). *Social Accounting and Economic Models*. Bowes & Bowes. London.

—— & D. A. ROWE (1958). Dynamic demand functions: some econometric results. *Economic Journal*, vol. 68.

THEIL, H. (1971). *Principles of Econometrics*. North-Holland, Amsterdam.

WAMPLER, R. H. (1970). A report on the accuracy of some widely used least squares computer programs. *Journal of the American Statistical Association*, vol. 65.

WIGLEY, K. J. (1968). R. STONE (ed.) *A Programme for Growth, Part 8: The Demand for Fuel 1948-1975: A Submodel for the British Fuel Economy*. Chapman & Hall, London, for Department of Applied Economics, University of Cambridge.

WOLD, H. (1953). *Demand Analysis*. Wiley, New York.

STATISTICAL SOURCES

ADMIRALTY. *Nautical Almanac and Astronomical Ephemeris*. HMSO, London, annually to 1959. Continued as *Astronomical Ephemeris*, from 1966 a publication of the Nautical Almanac Office (Royal Greenwich Observatory).

CENTRAL ELECTRICITY AUTHORITY. *Annual Report and Accounts*. House of Commons Papers, HMSO, London, annually to 1957.

CENTRAL STATISTICAL OFFICE:
> *Abstract of Regional Statistics.* HMSO, London, annually from 1965.
> *Monthly Digest of Statistics.* HMSO, London, monthly.
> *National Income and Expenditure.* HMSO, London, annually.

DEPARTMENT OF EMPLOYMENT AND PRODUCTIVITY (Department of Employment since 1970). Data from the Family Expenditure Surveys not published in the annual Reports. See also Ministry of Labour.

ELECTRICAL TIMES:
> *Electricity Supply Handbook.* The Electrical Times Ltd. London, annually.

ELECTRICITY BOARDS:
> *Annual Report and Accounts.* House of Commons Papers, HMSO, London, annually for each of the twelve Boards in England and Wales; annually to 1955 for the South East of Scotland and South West of Scotland Boards; and from 1955 for the South of Scotland Board.

ELECTRICITY COUNCIL :
> Data on file. The 1955 sample survey of domestic consumers. The 1961 sample survey of domestic consumers. The 1966 sample survey of domestic consumers. Sample surveys conducted by Audits of Great Britain Ltd, at 30 June 1963, 1964, 31 March 1965, 1966, 1967, 1968.
>
> *Utilization Research Reports,* numbers
>
> 7 The 1955 sample survey of domestic consumers, 1958;
> 42 The 1961 sample survey of domestic consumers;
> 55-58 Research into unrestricted domestic space-heating;
> 65 Grid load curve analysis 1965-66;
> 66 Research into unrestricted domestic water-heating;
> 75 The 1966 sample survey of domestic consumers, 1968.
> London.

GAS COUNCIL:
> Data on file.
> *Annual Report and Accounts*
> House of Commons Papers, HMSO, London, annually.

LONDON AND CAMBRIDGE ECONOMIC SERVICE. *The British Economy: Key Statistics 1900-66.* Times Newspapers Ltd, London, 1967.

METEOROLOGICAL OFFICE. *Monthly Weather Report of the Meteorological Office.* HMSO, London, monthly, with *Annual Summary, Introduction,* annually.

MINISTRY OF LABOUR. *Report of an Inquiry into Household Expenditure in 1953-54.* HMSO, London, 1957.
> *Family Expenditure Survey.* HMSO, London, annually from 1957, latterly as a publication of the Department of Employment and Productivity, and then the Department of Employment.
> *Ministry of Labour Gazette.* HMSO, London, monthly to 1968. Continued as Department of Employment and Productivity *Employment and Productivity Gazette,* and then as Department of Employment *Department of Employment Gazette.*

MINISTRY OF POWER (Ministry of Technology 1964-70, and thereafter Department of Trade and Industry)
> *Ministry of Power Statistical Digest.* HMSO, London, annually to 1967.
> *Digest of Energy Statistics.* HMSO, London, annually from 1969.

SOUTH WESTERN ELECTRICITY BOARD. Data on file.

INDEX

See also the List of Contents

aggregate tariff function defined, 131
Aitchison, J., 16n, 35n, 61n
appliance defined, 27
appliance ownership, correlation between
    exogenous variables and, 28, 34, 110
  data, 1, 4-5
  defined, 27, 32
  independence of utilization and, 28-9,
    33, 107
  predeterminateness of, 29, 108
appliances, classification of, 28, 33
  complementarity of fuel and, 8, 9,
    15-21, 26
  dynamics of adjustment to the stock of,
    8, 13-14, 122
  - heterogeneity of total stock of, 8-9, 13,
    23, 25
  homogeneity of classes of, 28, 33, 94,
    107
  independence of classes of, 28, 33, 107
  technical efficiency of, 7, 9, 23, 33, 50
arithmetic accuracy, 67, 69, 70
availability of gas, 1, 30, 34, 37-9, 52-3,
    83-4, 110-11
  coefficient of, 39, 91, 96-8, 122-3, 139-40

Badger, E. H. M., 64n
Bain, A. D., 5n
Balestra, P., 5n, 19-21, 24-5
Baxter, R. E., 22n
Brown, J. A. C., 6n, 16n, 22, 35n, 61n
Bulk Supply Tariff, 41n

Central Electricity Generating Board, 42
Cholesky decomposition, 69
Clean Air Act, 97
coefficients, interpretation of, 17-18, 32,
    35-7, 96-7, 116, 122-3, 140
  See also under individual variables
complementarity. *See* appliances,
    complementarity of fuel and
constant, inclusion of, 36-7, 110
  interpretation of, 37, 81
constraints on parameters, 5, 78, 81-4,
    89-91, 117-18
consumers, multi-household, 44
  number of domestic electricity, 1, 43-4, 48
consumption, domestic electricity. *See*
    quantity demanded of electricity

Cramer, J. S., 15-17, 26, 76n
Croft-Murray, G., 6n

Davies, M., 63n
daylight, hours of, 1, 5, 9, 14, 18, 26, 31,
    34, 47, 63-6, 77, 82-3, 98, 129
Deaton, A., 6n
degrees of freedom, 5, 79
demand for electricity defined, 8n
  See also quantity demanded of
    electricity
demand for fuel, total, 20, 21-3
dependent variable. *See* quantity
    demanded of electricity
disaggregation by type of appliance, 8-9,
    18, 24-5, 26
  over time, 8-10
Doumenis, G. J., 18-19, 24
Durbin, J., 75n, 76

Eastern Area, 60
Eastern Electricity Board's night and day
    tariff, 10n, 106
elasticities of demand defined, 95-6
  estimates of, 103-5
elasticities of utilization defined, 89-91
  best estimates of, 119, 139
Electrical Times, 41n
Electricity Board Areas, 1-2, 56
Electricity Council, 44, 51-2, 63, 113
error term, 29, 34, 36, 40, 46, 67-79, 108,
    111-12, 122, 134
errors in variables, 4-5, 34, 43-66, 74, 78-9,
    94-5, 97, 108, 113, 119, 130, 132, 139

exogenous variables defined, 32
expenditure, total consumption, 1, 4, 12,
    31, 34-5, 57-60, 62, 95-7, 109
  coefficient, 13, 35-6, 37, 59-60, 97, 104-5,
    122, 138, 140
extrapolation of time-series. *See* interpola-
    tion of missing values.

Family Expenditure Survey, 58
Farrell, M. J., 35n, 57n
Fisher, F. M., 20, 24-6
Forster, C. I. K., 22
Friedman, M., 57n

functional form of demand equation, 6, 8,
77
   of utilization functions. *See* utilization
   functions, form of

gas availability. *See* availability of gas
Gas Board Areas, 1-2, 56
Gauss-Jordan elimination, 69-70

habits, 8, 15, 18, 26, 61-3, 65, 129
hours of daylight. *See* daylight, hours of
household size and composition, 7, 9, 12,
   35, 103
   income effect and scale effect of, 36, 58
Houthakker, H. S., 13, 15, 17-18, 21-2, 24,
   26, 35n, 58n
Hurwicz, L., 75

illumination. *See* daylight, hours of
income, 6, 57, 62
index of the prices of all commodities
   except fuels. *See* prices of all commodities except fuels, index of
installed load, 27, 32
interpolation of missing values, 3-5, 34,
   48, 51, 74, 79, 95, 108
Ironmonger, D. S., 7n

Johnston, J., 71n, 77n, 79

Kaysen, C., 20, 24-6
Kleitman, N., 65n

lag operator defined, 31-2, 60-2
lags, 5-6, 8-9, 12, 17-18, 20, 26, 31-2, 34,
   43, 55, 60-3, 74, 85, 88-9, 95, 109, 117,
   128-9
lognormal distribution, 16-17, 35, 61-2
London and Cambridge Economic Service,
   59
London Area, 53, 60, 64
Longley, J. W., 69n, 71n
Lyness, F. K., 64n

Malinvaud, E, 8n, 68n, 78n, 141
Meek R. L., 41n
Merseyside and North Wales Area, 64
meter-reading, 44, 46-7, 94, 112
Ministry of Power, 21, 56
moving-quarter defined, 44
moving-quarter values, calculation of,
   46-7

natural variables, 7-9, 11, 26, 34, 63-6
   See also temperature, daylight, hours of
Needleman, L., 44, 59

normal weather conditions defined, 129
number of observations, 67, 113, 134

off-peak tariff defined, 106
   proportion of consumers on, 108
   See also price of electricity on off-peak
   tariffs
off-peak demand defined, 106
   See also quantity demanded of electricity
   on off-peak tariffs
omission of variables, 19, 24, 34-6, 66, 74,
   85, 94-5, 103
orthonormalization, 69
ownership. *See* appliance ownership

Pearce, I. F., 6n
Plackett, R. L., 69n
pooling of data, 5-6, 63
Prais, S. J., 35n, 58n
predetermined variables defined, 32
predictions, 6, 77-8, 101-3, 112, 125-6, 139
price function defined, 133
price of electricity on off-peak tariffs, 1,
   109-10, 114
   coefficient of, 5, 122
price of electricity on unrestricted tariffs,
   1, 6, 31, 34-5, 41, 53-5, 77, 109, 127-40
   and relationship to quantity supplied
   through tariff structure, 8-9, 13, 17, 19,
   24, 40-2, 127-40
   coefficient of, 5, 13, 37, 91, 97, 122,
   138-40
price of gas, 1, 31, 34-5, 37, 40, 55-7, 77,
   83-4, 109
   coefficient of, 5, 39, 91, 98, 104-5, 122,
   140
price of solid fuel, 1, 31, 34-5, 57, 62-3, 83,
   109
   coefficient of, 5, 37, 57, 59, 91, 97-8,
   122-3, 138, 140
prices of all commodities except fuels,
   index of, 6, 31, 60, 62-3, 95, 97, 109
prior information, use of, 78, 81
proportion of consumers owning defined, 32

quantity demanded of electricity, difference
   between actual and recorded, 29
   frequency distribution of, 131-2
   on off-peak tariffs, 1, 4, 9-10, 43, 107,
   113-14, 131
   on unrestricted tariffs, 9-10, 43-6, 48, 77
   total, 1, 3-4, 6, 43

Rees, R., 22n
Restricted Hour tariffs, 106
Rowe, D. A., 13-14, 20

Samuelson, P. A., 6n
seasonal pattern in consumption, 4, 14, 18, 94, 105, 131-2
Shell International Petroleum Co., 97n
significance level, 78-9
simultaneous equations models, 7, 9, 23, 25
Simultaneous Maximum Demand, 41
South Eastern Area, 60
South of Scotland Area, 6, 43, 77, 101-3, 112, 125-6, 139
South Wales Area, 3, 57
South Western Area, 3-4, 53, 94
Southern Area, 60
Standard Regions, 1-2, 58
Stone, 1, 6n, 11-14, 19-20, 24, 58n, 103-5
sunshine, hours of, 14
supply, 8-9, 17, 40-2, 111
    See also price of electricity on un-restricted tariffs, and relationship to quantity supplied through tariff structure

tariffs, electricity. See price of electricity on off-peak tariffs, price of electricity on unrestricted tariffs
tariffs, gas. See price of gas

tariff function defined, 53
tastes, 6-7
Taylor, L. D., 15, 21
temperature, 1, 4, 9, 12-14, 18, 26, 31, 34, 47, 63-6, 77, 82, 94, 109, 114, 129
    coefficient, 47, 66, 82, 98, 105, 123, 129
    threshold, 65, 84-5, 116-17, 122
Theil, H., 71n, 72n

U effect defined, 29, 81
utilization defined, 27-9
    estimates of, 85-7, 99-101, 117, 123, 139
    random element in off-peak, 111, 122
utilization functions, form of, 31, 85, 101-2, 109
    linearity of, 29, 32, 74, 108-9
    variables in, 31-2, 108-10

Wampler, R. H., 69n, 71n
wants, 7-9, 11
    See also tastes
Watson, G. S., 75n, 76
Whitting, I. J., 22
Wigley, K. J., 23-4, 26, 103-5
Wold, H., 6n

Z effect defined, 29, 81
Z-U effect defined, 33, 81